Solar, Wind and Lan

The global demand for clean, renewable energy has rapidly expanded in recent years and will likely continue to escalate in the decades to come. Wind and solar energy systems often require large quantities of land and airspace, so their growing presence is generating a diverse array of new and challenging land use conflicts. Wind turbines can create noise, disrupt views or radar systems, and threaten bird populations. Solar energy projects can cause glare effects, impact pristine wilderness areas, and deplete water resources. Developers must successfully navigate through these and myriad other land use conflicts to complete any renewable energy project. Policymakers are increasingly confronted with disputes over these issues and are searching for rules to effectively govern them. Tailoring innovative policies to address the unique conflicts that arise in the context of renewable energy development is crucial to ensuring that the law facilitates rather than impedes the continued growth of this important industry.

This book describes and analyzes the property and land use policy questions that most commonly arise in renewable energy development. Although it focuses primarily on issues that have arisen within the United States, the book's discussions of international policy differences and critiques of existing approaches make it a valuable resource for anyone exploring these issues in a professional setting anywhere in the world.

Troy A. Rule is an Associate Professor of Law at Arizona State University's Sandra Day O'Connor College of Law, USA. Prior to entering law teaching, he was an attorney at K&L Gates LLP in Seattle, Washington, USA, where his practice focused primarily on commercial real estate transactions and wind energy development.

Solar, Wind and Land

Conflicts in renewable energy development

Troy A. Rule

Routledge
Taylor & Francis Group

LONDON AND NEW YORK

First published 2014
by Routledge
2 Park Square, Milton Park, Abingdon, Oxon OX14 4RN

and by Routledge
711 Third Avenue, New York, NY 10017

Routledge is an imprint of the Taylor & Francis Group, an informa business

© 2014 Troy A. Rule

British Library Cataloguing-in-Publication Data
A catalogue record for this book is available from the British Library

Library of Congress Cataloging-in-Publication Data
Rule, Troy A.
Solar, wind and land : conflicts in renewable energy development / Troy A. Rule.
pages cm

ISBN 978-0-415-52046-1 (hardback) -- ISBN 978-0-415-52047-8 (pb) -- ISBN 978-1-315-77007-9 (ebook) 1. Energy development--Environmental aspects. 2. Renewable energy sources. 3. Energy policy--Environmental aspects. I. Title.
TD195.E49R85 2014
333.79'4--dc23
2014002645

ISBN: 978-0-415-52046-1 (hbk)
ISBN: 978-0-415-52047-8 (pbk)
ISBN: 978-1-315-77007-9 (ebk)

Typeset in Sabon by Fakenham Prepress Solutions, Fakenham, Norfolk, NR21 8NN

Printed and bound in Great Britain by
TJ International Ltd, Padstow, Cornwall

To Amy, my greatest friend

Contents

Illustrations

Figures

Tables

Acknowledgements

The origins of this book are ultimately traceable to Gary Hardke of Cannon Power Group and attorney Brian Knox of the Seattle, Washington, office of K&L Gates LLP, who entrusted me with significant real estate law matters relating to the development of the Windy Point/Windy Flats wind energy project in Klickitat County, Washington, despite my minimal experience at the time. My first hand exposure to the challenges and thrills of renewable energy development while working on Windy Point/Windy Flats first enticed me into this fascinating research area and continues to benefit me as a professor and scholar.

This book never would have been possible without generous support from the University of Missouri Summer Faculty Research Fund, the Robert L. Hawkins, Jr. Faculty Development Fellowship Fund, the Shook Hardy and Bacon Faculty/Student Scholarship Fund, the John K. Hulston General Fund, the Keith A. Birkes Faculty Research Fellowship Fund, and former dean Larry Dessem of the University of Missouri School of Law. Amanda Rhodes, Mike Powell, and Christine Bolton also provided invaluable research and editing assistance for the book. I am likewise grateful to Arizona State University's Sandra Day O'Connor College of Law, to Professor Brian Leiter of the University of Chicago Law School for helping me to initially break into the competitive world of legal academia, and to Professor Wilson Freyermuth of the University of Missouri School of Law for his friendship and mentoring throughout my time on the University of Missouri law faculty.

Abbreviations and acronyms

AWEA	American Wind Energy Association
BGEPA	Bald and Golden Eagle Protection Act
BLM	United States Bureau of Land Management
BPA	Bonneville Power Authority
CanWEA	Canadian Wind Energy Association
CSP	concentrating solar plant
DAHP	Department of Archeological and Historic Preservation
dBA	A-weighted decibels
DECC	United Kingdom Department of Energy and Climate Change
DOD	United States Department of Defense
DOE	United States Department of Energy
EHV	extra high-voltage
EIS	environmental impact statement
EOZ	energy overlay zone
EPAct 2005	Energy Policy Act of 2005
EPRI	Electric Power Research Institute
ESA	Endangered Species Act
FAA	United States Federal Aviation Administration
FCG	Friends of the Columbia Gorge
FERC	United States Federal Energy Regulatory Commission
FPL	Florida Power and Light
FWS	United States Fish and Wildlife Service
GW	gigawatt
Hz	Hertz
IEA	International Energy Agency
ITEDSA	Indian Tribal Development and Self-Determination Act
KPUD	Klickitat Public Utility District
kV	kilovolt
kW	kilowatt
kWh	kilowatt hour
LEED	Leadership in Energy and Environmental Design
MBTA	Migratory Bird Treaty Act

mph	miles per hour
MW	megawatt
NHPA	National Historic Preservation Act
NPH	Notice of Presumed Hazard
NWP	Northwest Wind Partners
PV	photovoltaic
RPS	renewable portfolio standard
SEPA	State Environmental Policy Act
SSCA	California Solar Shade Control Act
TERA	Tribal Energy Resource Agreement

Introduction

For more than a decade, wind and solar energy development have been advancing at a blistering pace throughout much of the world. Aggressive policies and rapid technological innovation have spurred an astonishing rise in renewable energy investment since the turn of the twenty-first century. In particular, solar panels and wind turbines are converting ordinary sunlight and wind into electricity with ever-increasing efficiency, helping to drive unprecedented growth in the wind and solar energy industries. From 2001 to 2012, the planet's installed wind energy generating capacity grew by more than tenfold.[1] Global increases in installed solar photovoltaic (PV) energy generating capacity have been even more dramatic, expanding by more than sixtyfold from 2000 to 2012.[2]

The world's recent turn to wind and sunlight for electric energy should come as no surprise, as both have served as important energy resources throughout much of human history in one form or another. Long before the Industrial Revolution, humans were harnessing wind currents to pump water, grind grain, and drive sailboats across great seas. And before the advent of light bulbs and automatic clothes dryers, windows and clothes lines enabled humans to illuminate their homes and to dry out clothing with solar radiation. Of course, modern renewable energy devices involve far more sophisticated uses of wind and solar energy, converting these natural resources into electric current—an incredibly fungible energy form that has improved living standards for billions of people all over the planet. As renewable energy technologies improve and the per-unit costs of generating electricity from wind and sunlight decrease, wind farms and solar arrays employing these technologies are growing ever more commonplace across the globe.

A growing body of evidence that fossil fuel emissions are warming our planet is further driving increases in global demand for wind and solar energy. Until recently, the world's electric power came almost exclusively from coal, nuclear energy, hydropower and natural gas—energy sources that tend to be neither as environmentally friendly nor as sustainable in the long run as sunlight and wind. As nations drill ever deeper underground for coal, oil, and natural gas and use ever more complicated methods to extract the earth's finite supplies of emissions-heavy fossil fuel resources, the advantages of harvesting renewable energy from wind and sunlight instead are ever more difficult to ignore.

Unfortunately, much more needs to be done if wind and sunlight are to ever reach their full potential as leading global energy sources. Enough solar energy strikes the surface of the earth every two hours to supply all of the world's energy needs for more than a year.[3] There is likewise enough energy in the planet's wind resources alone to meet the global energy demand.[4] Despite earth's abundance of wind and sunlight, these renewables still comprise only a small fraction of the global energy portfolio. And many of the most attractive renewable energy projects—those with excellent wind or solar resources, convenient transmission and road infrastructure, and few other obstacles—have already been built. The future growth rate of renewable energy thus depends in part on how easily developers can overcome the panoply of complex on-the-ground issues often involved in this unique type of development.

No amount of subsidies or tax credits can spur renewable energy development if irresolvable logistical obstacles get in its way. Land use conflicts can arise in nearly every type of renewable energy project, from modest rooftop solar installations to massive commercial wind farms. Solar panels are far less productive when shaded by neighbors' buildings or trees. Wind turbines are similarly less productive when spinning turbines situated immediately upwind of them create turbulent wakes that disrupt the currents flowing downwind. Wind and solar energy projects can encroach into endangered species' habitats, sacred cultural sites and even historic neighborhoods. They can require hundreds of miles of new transmission capacity through pristine, remote lands, and can disrupt the real-time operation of electric grids. And then there are the myriad disturbances that renewable energy devices can potentially inflict upon neighbors—turbine noise, flicker effects, radar interference, ice throws, fire hazards, stray voltage, glares, visual blight, and the list goes on and on.

Because the renewable energy sector is relatively young, legal rules to govern many of the unique land use conflicts associated with it have not been fully developed by legislatures or courts. Should landowners be liable for growing trees that shade a neighbor's solar panels? Should homeowner associations be allowed to adopt private covenants that prohibit renewable energy development within their communities? Should landowners be liable if their wind turbines disrupt the wind currents flowing into turbines on neighboring downwind parcels? And who should pay for the massive expansions of high-voltage transmission infrastructure that utility-scale renewable energy development often requires?

In the next half century, governments across the world will repeatedly be called upon to grapple with controversies associated with the global transition to more sustainable energy sources. What principles should govern decision making in these contexts? What analytical tools are best suited to assist policymakers in addressing these thorny, complex issues? The following chapters explore these questions and suggest some principles to guide policymakers in their effort to address the numerous conflicts inherent in renewable energy development without unnecessarily impeding the growth of this fledgling and critical industry.

Of course, developers cannot sit around and wait until all of the policy questions arising from renewable energy development are resolved before they begin pursuing these important projects. Until courts and legislatures have tailored laws to govern the full set of unique challenges associated with renewable energy development, developers and their legal counsel will have no choice but to navigate through some of these cloudy legal issues during the development process. In the midst of such uncertainty, a developer's or attorney's skill in anticipating and managing land use and other logistical issues can make all the difference between a project's failure and its success. Accordingly, this book seeks to be more than a policy book. It is intended as a way for renewable energy developers, utilities, affected landowners, and attorneys representing any of these parties to better educate themselves about the wide range of conflicts that can arise in wind and solar energy development and about some of the ways that these conflicts might be overcome.

Structure of the book

Chapter 1 of this book sets forth a policy foundation and theoretical framework for much of the analysis that follows as the book progresses. The chapter argues that the optimal set of policies governing renewable energy development is one that respects existing property rights while still maximizing the productive value of the scarce resources involved. The chapter also highlights the crucial distinction between airspace and the wind and sunlight that passes through it—a principle that resurfaces in multiple subsequent chapters.

Chapter 2 focuses on the broad array of conflicts that can arise between wind energy projects and more ordinary land uses on neighboring parcels. Among other things, the chapter discusses turbine "flicker" effects, wind turbine noise, claims of wind turbine syndrome, viewshed impacts, ice throws, and the property value impacts of wind farms. The chapter also describes conflicts between wind energy developers and military entities over the potential for commercial-scale wind turbines to disrupt military radar equipment. Comparing and contrasting various policy approaches that developers and local governments have used to prevent or ameliorate these potential problems, the chapter makes a handful of policy recommendations for better addressing these developer–neighbor clashes in future projects.

Chapter 3 describes disputes over wind turbine wake interference— conflicts that arise when wind turbines generate a "wake" of turbulent airflow that materially reduces the productive value of other turbines situated on neighboring downwind parcels. These sorts of conflicts are bound to become more prevalent as the density of wind energy development increases around the world. Should turbine wake interference be treated as an actionable nuisance? Or should landowners be entitled to "capture" any wind that crosses their properties without liability for downwind wake effects? Chapter 3 describes the policy approaches some governments have taken to address these conflicts and advocates a system of waivable setbacks as an alternative strategy. Such a system could potentially be a better means of reducing developers' uncertainty regarding wake effects while still encouraging developers to maximize the productive value of all wind resources involved.

Chapter 4 explores conflicts between renewable energy projects and wildlife, including bats, lizards, and avian species. Among other things, the chapter highlights recent disputes between wind energy developers and wildlife conservationists over proposed projects' threats to migratory bird and prairie chicken populations. The chapter also examines wildlife conservation issues associated with solar power plants, including a recent controversy over the potential impacts of a proposed solar power project on endangered tortoises in California's Mojave Desert. Illustrating the unique policy tensions between the competing goals of wildlife preservation and sustainable energy in these contexts, Chapter 4 describes some programs that have seemingly helped to reconcile these competing interests, including habitat mitigation lease requirements.

Chapter 5 discusses the "solar access" problem—the conflict that can arise when trees or structures on neighboring property shade a landowner's solar energy system and make it less productive. Uncertainty about the availability of undisturbed solar access over the useful life of a solar energy device can deter some landowners from investing in these systems. To address this problem, governments have enacted a diverse range of laws aimed at protecting solar access for landowners who install solar panels on their properties. These statutes were all enacted with good intentions, but some such laws are undeniably inefficient and incapable of effectively governing solar access disputes. This chapter compares and contrasts solar access protection laws and ultimately advocates a particular policy approach to these conflicts that has been adopted in at least one jurisdiction in the United States.

Chapter 6 focuses on the tensions that can arise when a renewable energy project threatens or degrades sacred tribal areas, historic buildings, or other precious cultural resources. A large percentage of the public lands in the Western United States that have abundant wind or solar resources also happen to hold historic or cultural value to one or more Indian tribes. This sort of cultural or historic value is tied to prime renewable energy lands in multiple other countries as well. Chapter 6 asks what policies and principles should guide stakeholders in determining what portions, if any, of these lands should be made available for renewable energy development. The chapter also examines controversies associated with rooftop solar panel installations on historic buildings and in historic districts. These disputes tend to raise similar questions about how to balance the competing ideals of renewable energy and cultural preservation.

Chapter 7 focuses on transmission-related obstacles to wind and solar energy development—some of the greatest impediments to the global movement to transition to more sustainable energy sources. In many parts of the world, the wind and solar resources with the greatest energy-generating potential are situated far from urban areas. Utilizing those resources thus necessitates the installation of expensive transmission infrastructure through multiple countries or local jurisdictions to deliver power to end-users. Who should pay for these new transmission lines, and who should have authority to determine where they are built? Another transmission-related issue relates to the intermittent nature of wind and solar energy resources. Because the wind is not always blowing and the sun does not always shine, the world's growing reliance on wind and solar energy can also create difficulties for grid operators. Wind and solar energy can complicate their task of constantly balancing electricity supply and demand on the electric grid, and significant debate exists as to how this problem should be addressed. Chapter 7 explores each of these challenges and asks what policies might most efficiently address them.

Chapter 8 provides an in-depth case study of how one renewable energy developer managed to overcome a laundry list of obstacles and controversies and develop a commercial-scale wind energy project—the Windy Point/Windy Flats project in Klickitat County, Washington, USA. In contrast to this book's other chapters, Chapter 8 does not focus on a specific category of conflicts affecting renewable energy development. Instead, it offers a close look at the development of a single wind farm, describing in detail the numerous challenges its developer faced and conquered to see the project to completion. The chapter offers a rare on-the-ground perspective on several of the topics covered in earlier chapters of the book, including wind turbine wake interference, transmission-related obstacles, conflicts with wildlife preservationists, and opposition from neighboring landowners concerned about viewsheds and other impacts.

Chapter 9 concludes the book by describing the role that technology and policy innovation could play in ameliorating many of the conflicts described in earlier chapters. Although the renewable energy movement is arguably still in its embryonic stages, much of the "low-hanging" fruit in global wind and solar energy resources has already been developed. To keep the movement charging forward, humankind will have to continue innovating at a rapid pace, finding ever more ways to access and capture the planet's renewable energy resources. Deep offshore wind energy, airborne wind turbines, and affordable energy storage are just some of the innovations with potential to grow dramatically in the coming decades. Of course, these new energy strategies are likely to trigger their own sets of unique and complex conflicts that policymakers and stakeholders will have to resolve. Chapter 9 offers a brief glimpse at some of these next-generation renewable energy strategies and discusses some of the new and perplexing conflicts they could raise. The chapter ends by emphasizing the important role that *policy* innovation can also play in facilitating the global adoption of renewable energy technologies as they evolve over time.

Because this book focuses exclusively on conflicts in *wind* and *solar* energy development, it intentionally omits discussions of numerous other emerging sustainable energy strategies that do not involve wind or solar resources. For instance, energy efficiency-focused policies such as green building laws, fuel efficiency standards, and green urban planning are often touted as some of the most cost-effective ways of reducing global carbon dioxide emissions. Likewise, alternative energy strategies such as fuel cells, biomass, small hydropower, wave energy, and geothermal energy are gaining importance as well. All of these promising strategies can obviously raise their own unique land use and logistical conflicts. Although such topics are certainly worthy of consideration in other venues, they are outside the intended scope of this book.

It is also worth stating at the outset that this book is not aimed at advocating *for* or *against* renewable energy development. Many of the topics it covers involve complicated webs of diverse impacts on wide ranges of stakeholders. In recognition of this fact, this book does not attempt to convince readers of the hazards of renewable energy development nor of the irrationality or extremism of those who oppose it. Industry and advocacy groups on both sides already provide a steady supply of material for readers wishing to focus on only one angle of these divisive debates. In contrast, this book seeks to offer a broader, more academic perspective on the clashes that wind and solar energy development inherently raise. Its descriptions of real-life battles over wind farms and solar arrays are intended to be enlightening and even entertaining for readers of all walks of life who are interested in these issues. The book can also serve as an educational resource for the countless stakeholders and policymakers tasked with managing the difficult trade-offs associated with renewable energy projects.

Notes

1 According to the Global Wind Energy Council, the global cumulative wind energy generating capacity had increased from 23,900 megawatts in 2001 to more than 282,500 megawatts by the end of 2012. *See Global Wind Report: Annual Market Update 2012*, Global Wind Energy Council (2013), p. 13, available at www.gwec.net/wp-content/uploads/2012/06/Annual_report_2012_LowRes.pdf (last visited June 3, 2013).
2 Specifically, cumulative global installed PV generating capacity has increased from a mere 1.5 gigawatts in 2000 to more than 96 gigawatts in 2012. *See Topic: Solar (PV and CSP)*, International Energy Agency website, available at www.iea.

org/topics/solarpvandcsp/ (last visited June 4, 2013); *A Snapshot of Global PV 1992–2012*, p. 4, International Energy Agency (2013).

3 The world's global primary energy demand in 2004 was approximately 480 exojoules (EJ). According to estimates by United States Department of Energy researchers, that amount of solar energy theoretically strikes the surface of the earth about every one and a half hours. Jeff Tsao, Nare Lewis, & George Crabtree, *Solar FAQs*, p. 10, United States Department of Energy (Working Draft version 2006, April 20), available at www.sandia.gov/~jytsao/Solar%20FAQs. pdf (last visited June 6, 2012).

4 *See* Kate Marvel et al., *Geophysical Limits to Global Wind Power*, 3 NATURE CLIMATE CHANGE 118, 120–21 (2012).

1 Economic and legal principles in renewable energy development policy

Many renewable energy enthusiasts have fantasized about a distant future day when our entire planet becomes fully energy-sustainable. In that utopian world, rooftops will be covered with photovoltaic (PV) solar shingles, wind turbines will grace the landscape like wildflowers, and all cars will run on batteries powered by the wind and sun. Coal smokestacks will no longer mar the skyline, and the air and water will be clean and clear.

In reality, the earth has a very long way to go before it approaches anything close to true energy sustainability and getting there will require countless trade-offs. Wind turbines can kill birds and alter pristine viewsheds. Solar panel installations can trigger neighbor disputes over shading and disrupt neighborhood aesthetics. And many sustainable energy strategies can complicate the management and funding of electric grids.

In other words, renewable energy projects do more than just generate clean, sustainable energy; they can also impose real costs on the communities that host them. Ineffective handling of these costs can spell the demise of a proposed renewable energy project, even if its aggregate benefits to society would easily outweigh its costs.

Savvy renewable energy developers are skilled at anticipating the costs that their projects might impose on third parties. They have strategies for mitigating these costs and for preventing them from unnecessarily thwarting a project's success. The policymakers that most effectively serve their electorates' interests in the context of renewable energy development are similarly adept at recognizing its potential costs and benefits and at structuring laws accordingly.

Humankind can only achieve a fully sustainable global energy system if the countless stakeholders involved in renewable energy development work together to resolve the myriad property conflicts associated with it. What principles should guide developers and policymakers in their ongoing efforts to prevent these inevitable conflicts from impeding the growth of efficient, equitable renewable energy development across the world? Serving as a foundation for the other chapters of this book, this chapter highlights some general concepts that can aid developers, policymakers, and other interested parties as they strive to manage the complex web of competing interests that often underlies modern renewable energy projects.

The microeconomics of renewable energy policy

Basic microeconomic theory can provide a useful framework for describing the incentives of the various players involved in renewable energy development. Simple economics can also bring the overarching objectives of renewable energy policies into focus in ways that are not possible through any other analytical tool. Because of economic theory's unique capacity to elucidatethe complicated trade-offs at issue, anyone tasked with structuring laws to govern renewable energy should have a solid grounding in it.

Viewed through an economic lens, the optimal set of policies to govern renewable energy development is one that ensures that society extracts as much value as possible out of its scarce supply of resources. A given renewable energy project best promotes the social welfare—the overall well-being of all of society—only when it constitutes the highest valued use of the land and airspace involved. More broadly, social utility can be maximized only when humankind develops renewable energy projects solely on land and in airspace for which there is no other alternative use that is worth more to society.

optimal = highest valued use of the land

Unfortunately, discerning whether a renewable energy project is the highest valued use of land and airspace at a given location is anything but easy. Just as no two parcels of land are identical, no two parcels are equally suitable for renewable energy. Although wind and sunlight can be harvested to some extent almost anywhere on earth, certain regions and places naturally have far more productive wind or solar resources than others. As illustrated by wind and solar resource maps available on the Internet,[1] the potential energy productivity of wind and solar resources varies dramatically across the planet. Those variations can significantly impact whether renewable energy development is optimal in a given location.

potential E productivity

In addition to variations in the quality of the wind or solar resources, countless other site-specific factors can affect the desirability of development at a given site. From a commercial wind energy developer's perspective, the availability of road access, transmission capacity, and "off-takers" interested in purchasing a proposed project's power at a favorable rate can greatly influence whether a site is well suited for development. In the context of rooftop solar development, an existing building's orientation toward the sun and the slope and shape of its roof can impact whether installing a solar PV system on the building makes good economic sense.

site-spe-cific factors

Importantly, even many locations with exceptional wind or solar resources and favorable site-specific factors like those just described are nonetheless poor venues for renewable energy development. Specifically, projects in some of these locations may impose costs on neighbors and other stakeholders that are so great that they exceed the project's benefits. These "external" costs—costs not borne by the primary participants in the development process—are largely the focus of this book.

external costs

Positive externality problems

When the parties directly involved in renewable energy development ignore the external costs and benefits associated with their projects, market failures that economists describe as "externality problems" often result.[2] Basic economics principles teach that some sort of government intervention—a new legal rule or policy program—is sometimes the best way to mitigate externality problems and thereby promote greater economic efficiency.

In recent years, policymakers around the world have made great strides in crafting programs to address the *positive externality problems* plaguing the renewable energy sector. Positive externality problems exist in this context because wind and solar energy development generate significant *benefits* that are distributed diffusely among the planet's billions of inhabitants, most of whom reside far away from the project site. Such benefits include reductions in the global consumption rate of fossil fuel energy sources and consequent reductions in harmful emissions, including carbon dioxide emissions that may contribute to global warming. Among countries that are net importers of fossil fuels, increasing the proportion of the national energy demand supplied through renewable resources can also promote economic growth, improve trade balances and advance greater economic stability.

Without some form of government intervention, rationally self-interested landowners and developers tend not to adequately account for external benefits in their decisions and thus engage in sub-optimally low levels of renewable energy development. Consider, for example, a rural landowner who leases thousands of acres of land to a developer for a commercial wind farm project.[3] Such lease agreements typically give landowners an estimate of the monetary compensation they can expect to receive if the developer's project is completed. Through power purchase agreements or other means, developers can also approximate how much financial compensation they will earn in connection with a given project. By weighing the expected benefits that would accrue directly to them against their own budgetary or other costs, landowners and developers can make rational, self-interested decisions about their involvement in the wind farm.

However, as mentioned above, renewable energy projects also generate numerous *other* benefits that accrue more generally to thousands or even billions of other people. In a completely free and open market, these ancillary benefits are not generally captured by project landowners or developers. Rational, self-interested landowners and developers are thus likely to place little or no value on these outside benefits and will instead weigh only their *own* potential benefits and costs when making development-related decisions. Because they rationally choose to ignore these external benefits, developers and landowners are likely to engage in sub-optimally low levels of renewable energy development without some form of government intervention.

Figure 1.1 below depicts this concept graphically using ~~simple marginal cost and marginal benefit curves.~~ Suppose that, for a given wind or solar energy development project, a developer's marginal cost of installing additional units of renewable energy generating capacity appears as curve MC. The tendency for developers to focus solely on their own marginal benefits from additional renewable energy installations (represented by the marginal cost curve MB_p) rather than on the broader social benefits of such additional capacity (MB_s) can lead them to install a quantity of installed capacity (Q_1) that is less than the socially optimal amount (Q^*). ~~In summary, unless governments intervene to correct them, positive externality problems can result in an inefficiently low quantity of renewable energy development.~~

Governments across the world are well aware of this positive externality problem associated with renewable energy and have formulated a wide array of creative programs and policies aimed at countering it. Many of these programs ~~seek to reduce the market cost of renewable energy development through direct subsidies, investment tax credits, special financing programs, streamlined permitting processes, and related means so as to effectively shift developers' marginal cost curves downward.~~

This intended effect of incentive programs on the pace of renewable energy development is illustrated in Figure 1.2. Government subsidies, investment tax credits and similar cost-reduction programs ~~seek to shift the marginal cost curve for renewable energy development downward from MC to MC_i.~~ All else equal, such cost reductions incentivize landowners and developers to increase the quantity of installed renewable energy generating capacity from Q_1 to Q^*.

[handwritten margin note: gov soluhons in place]

[handwritten note: goal: reducing marginal cost curve (↓costs for landowners & developers)]

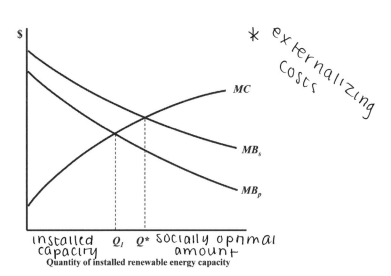
*[handwritten note: * externalizing costs]*

Figure 1.1 Positive externality problems in renewable energy development

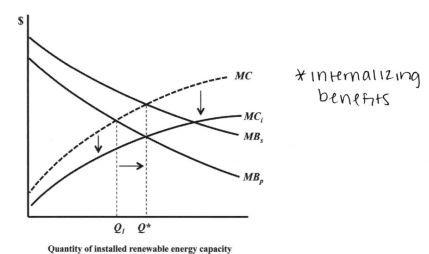

Figure 1.2 Intended effect of subsidies and related cost reduction programs on the quantity of renewable energy development

In contrast, ~~renewable portfolio standards, feed-in tariffs and some other incentive programs seek to correct or mitigate the positive externality problems associated with renewable energy by helping developers to internalize more of the external benefits of their developments.~~ These programs, which directly or indirectly increase the market price of renewable energy, effectively shift developers' private marginal benefit curve (MB_p) outward so that it matches or more closely approximates the social marginal benefit curve (MB_s). Much of the recent growth in the global renewable energy sector can be attributed to these types of policies.

~~Renewable portfolio standard (RPS) programs~~ enacted in much of the United States exemplify this policy strategy. These programs, which typically require utilities in affected jurisdictions to purchase some minimum amount of their energy from renewable sources, ~~increase the market demand for and price of renewable energy.~~[4] The United States' per-kilowatt "production tax credit" for various types of renewable energy[5] and various "feed-in tariff" programs common in some other countries[6] are even more direct versions of this approach. ~~Production tax credits and feed-in tariffs increase the marginal benefits of renewable energy development by artificially increasing the sale price of renewable energy itself.~~

As shown in Figure 1.3 below, to perfectly correct for positive externalities associated with renewable energy development, a policy strategy of relying solely on one of the aforementioned programs would ~~need to be calibrated to shift developers' private marginal benefit curve associated with development from MB_p to MB_{pi} such that it matched the social marginal benefit curve (MB_s).~~ If successful, such a program would thereby increase the quantity of installed renewable energy capacity to Q^*, the optimal level.

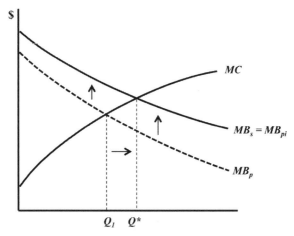

Figure 1.3 Intended effect of RPS programs and feed-in-tariffs on the quantity of renewable energy development

Negative externality problems

Policymakers and academics devote comparatively less attention to many of the *negative externality problems* associated with renewable energy development. These problems arise when some of the costs associated with renewable energy projects are borne by individuals who do not directly participate in and have relatively less influence over development decisions.

The list of wind and solar energy development's potential costs to outsiders is expansive and continues to grow. Within the land use realm alone, this list includes aesthetic degradation, noise, ice throws, flicker effects, interference with electromagnetic signals, destruction of wildlife habitats and wetlands, bird and bat casualties, disruption of sacred burial grounds or historical sites, inner ear or sleeping problems, annoyance during the construction phase, interference with oil or mineral extraction, solar panel glare effects, exploitation of scarce water supplies, and heightened electrical, lightning, and fire risks. Theoretically, an excessive quantity of renewable energy development can result when project developers base their renewable energy development decisions solely on their own anticipated costs and ignore costs borne by others.

It is true that most of the costs of renewable energy development are borne by developers and by the landowners on whose property the development occurs. Developers typically incur the greatest costs associated with any given renewable energy project. They frequently invest large sums of money in project planning, buying or leasing real property, paying for adequate road and transmission infrastructure, purchasing expensive

[margin annotations:] costs imposed on 3rd parties · examples · costs for developers

renewable energy systems, navigating the often-confusing government-permitting scheme applicable to the project, and ultimately installing the systems. Landowners who lease or otherwise convey interests in their real property for such development can also incur significant costs associated with the presence of renewable energy devices and related infrastructure on their property. These costs borne by developers and landowners—the primary decision makers in wind or solar energy projects—are represented by MC_p in Figure 1.4 below.

However, countless other parties, including those who own or use neighboring land, can also suffer significant costs associated with new wind or solar energy installations. These neighbors and other affected outsiders often have limited influence on where and how projects are built, and rational self-interested developers and landowners do not fully account for neighbors' costs when making development decisions. This classic *negative* externality problem can theoretically lead to an excessive amount of renewable energy generating capacity. As shown in Figure 1.4, when developers consider only their own costs associated with installing each incremental unit of renewable energy generating capacity in their projects (MC_p), rather than considering the aggregate social cost of such incremental units (MC_s), excessive levels of development (Q_2) can result.

More specifically, without government intervention, developers may install renewable energy facilities in unjustifiable locations simply because they do not bear all of the costs associated with doing so. Wind farms may hazardously encroach into migratory bird paths, solar energy projects may

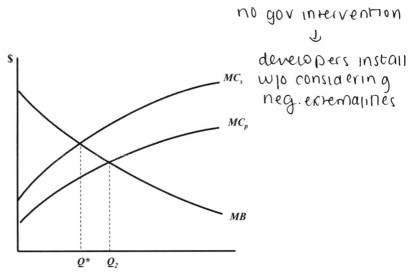

Quantity of installed renewable energy capacity

Figure 1.4 Effect of *negative* externalities on quantity of renewable energy development

disrupt sacred Native American burial sites, or numerous other avoidable conflicts may result. Such conflicts are not only inefficient; they can also tarnish renewable energy's public image in ways that ultimately impede global progress toward energy sustainability.

When policymakers fail to anticipate and address the negative externalities associated with wind and solar energy development, outside parties who suffer injuries as a result become more likely to mobilize and oppose future projects—even worthwhile ones that pose minimal risks. And when laws governing renewable energy land use conflicts are unsettled and unclear, the resulting uncertainty can also slow the pace of development. Accordingly, any comprehensive set of policies for promoting renewable energy development must do more than merely correct positive externality problems by incentivizing more of it. It must also appropriately account for the external costs and conflicts that can arise from these projects.

Cost and benefit distributions in renewable energy development

Of course, the simplistic marginal cost and benefit curves shown above do not paint a full picture of the complex dynamic that underlies most renewable energy development projects. Almost every such project creates some winners and some losers. The parties affected by any proposed renewable energy development and their respective levels of support inevitably vary but are worth considering in connection with any major project.

The diagram in Figure 1.5 illustrates the basic distribution of support for a typical proposed renewable energy project. The small, centermost circle in the diagram represents those parties proposing it. In some cases, such as when a homeowner purchases and installs a solar array on her rooftop or a small wind turbine in her backyard, the only party in this circle is owner of the project site. In other cases, such as when a property owner leases land for a wind farm project in exchange for a promise of payments upon the project's completion, a developer and one or more separate landowners are in this circle. Regardless, the parties in this centermost circle are typically supportive of the project and believe that it will confer a positive net benefit to them; otherwise, they likely would not be proposing it in the first place.

The parties in the outermost circle in Figure 1.5 are those that have no direct interest at stake in a proposed renewable energy project. These parties are also likely to support the project or to abstain from actively opposing it. Non-stakeholders living hundreds of miles away from a proposed wind or solar project have little reason to expend time or effort in trying to stop it. Renewable energy development produces several well-established global environmental benefits that often extend far beyond the immediate vicinity of a given project to such parties. These generic benefits help to explain the broad general support that the wind and solar energy industries enjoy throughout much of the world.[7] They also help to explain why many national governments have sought to encourage renewables through subsidies and other non-site-specific incentive programs.

The greatest variation in support for a particular renewable energy project exists among those in the shaded middle circle in Figure 1.5. This middle circle tends to be populated by neighbors, environmental or wildlife advocacy groups, and other parties other than a project's developer or landowners. The parties in this middle circle generally do not receive direct financial benefits if the project is completed, yet they perceive that the project will significantly impact them or their interests.

Nearly all of the land use conflicts described in this book involve the external stakeholder parties represented by this shaded middle circle. The success or failure of any proposed renewable energy project hinges in part upon how well its developer and policymakers in the host jurisdiction address the conflicts involving these individuals and entities. Whenever a proposed renewable energy project's potential net benefits to those in the innermost and outermost circles in Figure 1.5 would exceed its net costs to those in the shaded middle circle, developing the project promotes the net social welfare. Inefficiency results in these contexts whenever opposition from those in the middle circle prevents such a net-welfare-maximizing project from proceeding.

Of course, even this characterization of the costs and benefits involved in renewable energy development is a bit oversimplified. Specifically, it fails

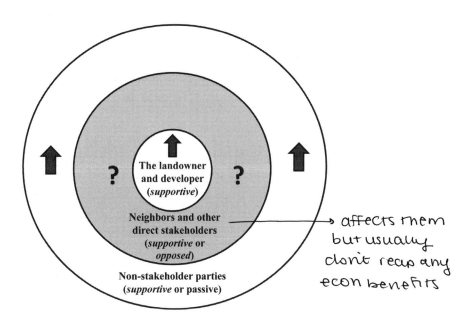

Figure 1.5 Typical distribution of support for a proposed renewable energy project

to distinguish costs and benefits affected by the specific location and layout of a proposed project from those costs and benefits that are largely uninfluenced by the nature or location of the project site.

Consider, for instance, the benefits that can accrue to wind turbine or solar PV manufacturers in connection with a large renewable energy project. These parties can have a substantial financial interest in the successful development of a proposed major wind farm or solar PV installation. If the project receives approval and proceeds, the manufacturer will sell valuable inventory and generate income. However, such parties' interests deserve relatively little weight in renewable energy siting decisions because they are rarely site-dependent. A wind turbine manufacturer receives roughly the same net profit for each turbine it sells and is thus likely to provide about the same level of support for any wind energy project that would use its turbines, regardless of the project's location. The potential gains to these manufacturers are often at least partly accounted for in their sale negotiations with developers.

Likewise, the potential losses to large coal and natural gas companies from reductions in product demand caused by renewable energy projects typically are not site-specific either. Their opposition to particular projects thus arguably also warrants somewhat less local government attention in project siting decisions. Consideration of the interests of these sorts of non-site-specific stakeholders, listed in the left column of Table 1.1, seems more appropriate in the context of national policymaking than in local siting analyses.

In contrast, the interests of many other non-developer stakeholders are very site-specific and thus highly relevant in the renewable energy siting process. For instance, the potential impacts of a proposed wind farm on neighboring landowners or on local wildlife species deserve comparatively more consideration in siting decisions than the impacts on the manufacturer that would supply the turbines. The same is true for indigenous groups

Table 1.1 Types of non-developer stakeholders in renewable energy projects

Stakeholders with non-site-specific interests	Stakeholders with site-specific interests
Turbine or PV manufacturers	Neighbors
Fossil fuel producers	Conservation groups
Renewable energy advocates	Wildlife advocacy groups
	Local indigenous groups
	Historic preservation advocates
	Local utility companies

who worry about how a particular proposed project might disturb their ancestral lands. Conflicts involving these sorts of stakeholder parties, listed in the right column of Table 1.1, are the primary focus of this book.

Land and airspace as renewable energy property

In addition to comprehending the complex trade-offs inherent in renewable energy siting decisions, effective policymakers must also have a clear understanding of the specific types of resources that are at issue in these contexts. One additional background principle that recurs multiple times in this book is that wind currents and sunlight themselves are *not* private property. At times, policymakers worldwide have overlooked this fact in their zealous effort to promote wind and solar energy development throughout the globe. Recognizing the distinction between airspace and the wind and sunlight that flow through it is thus critical to effectively governing property conflicts in the context of wind and solar power.

It is difficult to accurately analyze many land use conflicts over renewable energy development without a clear understanding of the property interests at stake in these disputes. Over the past quarter century, the growing importance of wind and solar energy has led some to advocate the legal recognition of new types of property interests such as "wind estates"[8] or "solar rights."[9] In reality, these supposedly newfound property types are often just re-characterizations of existing property rights that laws have clearly defined for centuries. Hoping to counter this trend, this book advocates a more consistent approach to analyzing clashes over wind and solar energy development, framing the property rights involved solely as interests in *land* and in the *airspace* immediately above it.

Property in sunlight and wind?

The renewable energy movement has significantly altered the way that humans view wind and sunlight. Although humans throughout history have appreciated sunny days and cool summer breezes, sunlight and wind have not historically been viewed as legally protected commodities capable of being bought or sold. However, due to a variety of factors, global attitudes about the marketability of wind and sunlight are gradually changing. As modern technologies make it ever easier to convert wind and solar radiation into electricity, these unique natural energy sources are increasingly seen as having substantial economic value and therefore being worthy of commodification.

Viewed outside of the renewable energy context, the notion that wind and sunlight might be marketable property assets seems absurd. After all, neither wind nor sunlight is particularly scarce. Both are present throughout much of the day almost everywhere on earth. Basic economics teaches that something must be scarce to command a market price. Why, then, are countless private individuals throughout the world today receiving large sums of money under wind and solar leases and similar arrangements?

The not-so-obvious answer to this question is that developers who sign wind and solar energy leases are not really paying for wind or sunlight. The assets truly transferred under such agreements are carefully-defined interests in land and in the airspace just above its surface—assets that are inherently scarce. Installed wind turbines occupy surface land and extend high into airspace, precluding any other physical occupation of the space they inhabit and limiting the range of viable uses of land and airspace in their immediate vicinity. Similarly, solar energy systems reside upon rooftops and other surfaces and require that some of the airspace near them be kept open to prevent shading. Renewable energy developers secure property rights in surface land, rooftops, and airspace—not wind and sunlight—to develop their projects. Thus, the emergence of renewable energy technologies over the past few decades does not warrant legal recognition of some new property type. Laws in most jurisdictions have clearly allocated private and public property interests in the land and airspace resources involved in wind and solar energy development for hundreds of years.

Viewing renewable energy development conflicts as disputes over land and airspace rather than as disputes over wind or sunlight can alter the analysis of these disputes in ways that sometimes frustrate some renewable energy advocates. Such frustration arises because this paradigm precludes policy-makers from ignoring existing land and airspace rights in their eagerness to promote renewable energy. Renewable energy development is merely one of several potential ways of using any given set of land and airspace resources, and in many cases it is not the most efficient or productive of all conceivable uses. Commercial wind and solar energy projects can alter the character of vast stretches of land and airspace, and materially change the appearance and culture of communities. Even a fervent supporter of wind energy would not advocate replacing Paris' iconic Eiffel Tower with a wind turbine. Nor would it make sense for New York City officials to limit the heights of all new downtown buildings to just three stories solely to protect sunlight access and encourage rooftop solar panel installations. What policies, then, should govern public and private decisions over which land and airspace should be dedicated to renewable energy generation and which should be reserved for other uses?

The best possible rules for addressing renewable energy land use conflicts are those that, all else equal, are most likely to channel the scarce land and airspace involved in such conflicts to their highest valued uses. The balance of this chapter discusses land and airspace rights in the context of renewable energy development in greater detail, elaborating in particular on the growing role of airspace in the sustainability movement and the potential implications of this trend.

Renewable energy development as a use of land

In a sense, renewable energy development is ultimately just another form of real estate development. Like most commercial real estate projects, utility-scale wind and solar energy projects essentially involve the construction of large structures upon the surface of land for commercial purposes. It should come as no surprise, then, that many of the same basic property law issues considered during the early stages of commercial real estate developments are relevant in the renewable energy context as well. For instance, is the seller or lessor of the subject property its sole owner in fee simple? Are there any financial liens or encumbrances on title? Will the developer have adequate road and utilities access, or are additional appurtenant easements needed? Could any of the covenants, servient easements, or other encumbrances recorded against title potentially interfere with the development plan? These sorts of generic real estate development questions, while not the focus of this book, are just as capable of derailing a wind or solar energy project as they are at thwarting progress on any other type of real estate development.

On the other hand, the unique characteristics and enormous size of utility-scale renewable energy projects expose them to a whole host of *additional* land use issues that rarely or never arise in the development of an ordinary shopping center or condominium. A single wind farm can involve dozens of square miles of contiguous land. Modern commercial wind turbines frequently stand more than 400 feet high and must be spaced great distances apart to minimize turbine wake interference.[10] Turbines also must be spaced sufficiently far from property boundary lines and roads so that they would not fall onto neighboring parcels if a catastrophic event toppled them over. Although concentrating solar plants (CSPs) tend not to be as towering or sprawling as wind farms, they can also occupy substantial amounts of surface land and raise similar sorts of issues.

normal real estate issues + some unique ones

Renewable energy development as a use of airspace

Conflicts over competing uses of surface land are relatively straightforward, but clashes over the *other* natural resource at the center of many renewable energy development conflicts—airspace—can be far more perplexing. As the following article excerpt describes, the sustainability movement is making airspace more valuable than ever before and further complicating efforts to govern this important resource.

> Airspace is among the most ubiquitous of all natural resources, present in every corner of the globe. Nonetheless, airspace is inherently scarce. Each cubic inch of it exclusively occupies a unique spatial position in the universe. The old adage "location, location, location" thus applies as much to the valuation of airspace as it does to the valuation of surface

land: ownership rights in a cube of remote, high-altitude airspace might be worth only pennies, even though rights in an equivalent volume of airspace above a city's downtown core might fetch millions of dollars.

Airspace is distinct from "air"—the life-sustaining blend of mostly nitrogen and oxygen gases that circulates around the planet. Because air pollutants freely course throughout the world's air supply, *air* is sometimes characterized as a globally-shared "commons." In contrast, much of the *space* through which air flows is not a commons but is separately owned or controlled.

Similarly, airspace is distinct from the countless invisible waves that pass through air. Vibrating objects transmit electromagnetic waves of varying frequencies through the air to deliver music, spoken words, and other sounds to our ears. Modern electronic equipment can also transmit waves of varying frequencies through air, including waves capable of communicating information via devices such as cellular phones, radios, and wireless computer receivers. The radio spectrum itself is a highly regulated commons in the United States, subject to detailed policies from the Federal Communications Commission for allocating transmission rights at various frequencies among private and public parties. However, all of these waves and transmission frequencies are distinct from the airspace through which they commonly pass.

On a similar theory, airspace is also fully distinguishable from the wind currents and solar rays that fuel renewable energy generation. Airspace can be largely devoid of wind and sunlight on calm evenings and yet serve as the medium through which those resources travel on blustery days. Because wind and solar radiation are practically inexhaustible, they arguably warrant no private property protection under conventional property theory. In contrast, airspace is a finite, immovable resource that has justifiably enjoyed property protection for centuries … Airspace laws originally evolved much as classical property theorists might have predicted: through the emergence of property rights in what was a previously shared commons in response to technological innovation.

For much of recorded history, most of the earth's airspace was beyond the physical reach of humankind so few conflicts could arise regarding its use. Out of practical necessity, the majority of the planet's airspace was merely a commons through which landowners enjoyed sunlight and views. Laws in ancient Rome recognized that surface owners held rights in the airspace above their land. The "doctrine of ancient lights" at English common law also indirectly limited some building heights in order to protect neighbors' access to sunlight. However, in early, agriculturally-based societies, most landowners were primarily concerned with having rights in enough airspace to enable their crops to grow.

As construction techniques gradually improved over the centuries, airspace became an increasingly valuable resource and rules clarifying property interests in airspace naturally followed. Legal historians have traced the beginnings of modern airspace law as far back as the 1300s, when Cino da Pistoia pronounced the maxim: *Cujus est solum, ejus est usque ad coelum*, or "[to] whomsoever the soil belongs, he owns also to the sky." This doctrine, commonly known as the "*ad coelum* rule," established simple private property rights in airspace based upon subadjacent parcel boundaries. The rule was subsequently cited in Edward Coke's influential commentaries in the seventeenth century and in William Blackstone's commentaries in the eighteenth century, cementing it as a fixture in English and American common law. By the early twentieth century, U.S. courts were applying the doctrine to find trespass for even minor intrusions into the space above privately-owned land. [...]

[N]ew airspace uses emerging from the sustainability movement in recent years have further complicated the task of governing airspace rights. Difficult new policy questions are arising in part because ... some sustainable land use strategies require the occupation of additional airspace while other strategies necessitate that more airspace be kept open.

Growing calls for open airspace

Several types of renewable energy and green development strategies require open airspace. Unoccupied airspace allows sunlight and wind to reach plants, buildings, and renewable energy devices without interruption—a valuable function in our increasingly green economy.

One sustainability-oriented use for open airspace is to prevent the shading of solar energy devices. The concept of guarding against solar panel shading is commonly known as "solar access" protection and has been the subject of numerous statutes, ordinances, and law review articles over the years. Photovoltaic solar panels and passive solar energy strategies generate and save significant amounts of power, thereby reducing an economy's dependence on fossil fuels and other conventional energy sources. However, these and most other solar energy devices are far more productive when exposed to direct sunlight. Shade from trees or other structures in the airspace above nearby land can thus diminish a solar panel's productivity. The risk that trees or buildings could ultimately pop up in neighboring airspace and shade solar energy systems deters some landowners from investing in rooftop solar installations.

The demand for laws to protect solar access has rapidly grown over the past few years due to an unprecedented interest in rooftop solar energy development. Accordingly, government-provided incentive programs are supporting solar energy more aggressively today than ever before. As the cost-effectiveness of rooftop solar energy grows, so does the need for laws enabling landowners to protect their solar energy systems from shading by neighbors.

Urban gardens can also require a degree of direct solar access that only open city airspace provides. An increasing number of cities throughout the country are encouraging the cultivation of gardens on inner city lots as a means of combating blight and improving urbanites' access to fresh local produce. Of course, the successful growth of many food plants requires adequate sunlight. This need for unobstructed sunlight can also potentially constrain the development of airspace above neighboring parcels.

Even in the context of green building, access to sunlight via open airspace has taken on added value in recent years. The United States Green Building Council's 2009 Rating System rewards points toward LEED (Leadership in Energy and Environmental Design) Certification for building designs that satisfy specific natural daylight illumination requirements. Natural lighting through skylights and windows conserves energy by mitigating the need for electrical light. Unfortunately, shade from neighboring buildings or trees can reduce the degree of interior illumination achievable on a given parcel, necessitating greater reliance on electricity-dependent artificial light sources.

One other potential use for unoccupied urban airspace is to provide wind access for "small" wind turbines. Wind turbines convert the kinetic energy in wind into electric power. Small turbines are petite versions of commercial turbines and usually generate only enough power to offset a portion of a single landowner's energy consumption. These devices may not require direct sunlight, but they still need a substantial amount of open airspace to function effectively. They are more productive the higher they reach into the sky and can require that hundreds of feet of open airspace in the upwind direction to ensure that winds flowing into them are largely undisturbed. Wind access for small turbines has historically been a low priority in most jurisdictions, although this could change as these devices become increasingly cost-effective and demand for them continues to grow.

Mounting pressure to occupy airspace

Despite the myriad uses for *open* airspace just described, physically occupying airspace with buildings, trees, or other structures also creates substantial value in many circumstances. The sustainability movement is also bolstering demand for this other category of airspace uses.

Filling more urban airspace with buildings has become an increasingly attractive policy option in the past few decades as the damaging effects of suburban sprawl have become evident. Sprawling development on the suburban fringe tends to require more public infrastructure than dense urban in-fill projects and can also result in comparatively longer commutes and greater energy consumption. As a result, some have advocated land use policies that loosen height restrictions and make more intense use of urban airspace. Taller buildings provide more work and living space per acre, leaving more land available for green spaces and easing the pressure for suburban expansion. High-rise development can be particularly valuable when situated near public transit systems because of the comparatively low burden it places on traffic congestion and transportation infrastructure. For these and other reasons, vertical building designs are commonly viewed as relatively eco-friendly approaches to real estate development.

Cities are also seeking to enhance their environmental sustainability by filling more of their skylines with trees. In recent years, tree preservation and planting programs have begun sprouting up in major cities throughout the world. In some regions, trees that shade buildings can reduce air conditioning usage during hot summer months. Trees can also help to improve stormwater drainage, sequester carbon dioxide emissions from the air, and improve the aesthetic ambiance of city streets. In certain circumstances, trees can even reduce heating energy costs on cold days by serving as windbreaks for homes and other buildings. Of course, trees must physically occupy scarce airspace to perform these valuable functions.

Unlike buildings and trees, small wind turbines are a relatively new type of structure competing to occupy urban airspace. Small wind turbines not only require open airspace for adequate wind flow, but also fill substantial airspace with their towers and rotor blades. Although wind energy devices have historically been installed primarily in rural areas, permit applications to install small wind-turbines in suburban areas with heights upwards of 120 feet are increasingly common.

(Troy A. Rule, *Airspace in a Green Economy*. 90 UCLA L. Rev. 270, 274–79 & 285–90 (2011)[11])

In addition to the trees, buildings, and small wind turbines described in the preceding excerpt, commercial-scale wind turbines are also increasingly occupying large amounts of valuable airspace and inevitably impacting neighboring land uses. This growth in the importance of airspace highlighted above is likely to only increase in the coming years. In the midst of these changes, policymakers around the world face the daunting task of crafting legal rules to govern renewable energy development in ways that promote optimal use of the planet's finite supply of airspace and land.

From general to specific

Having introduced several economic and legal concepts that are generally applicable in the context of renewable energy development, this book now proceeds by examining specific types of conflicts that can arise. Many of the principles outlined in this chapter will reappear and help to shape the analysis of particular conflicts in the chapters that follow.

Notes

1 *See, e.g.,* http://www1.eere.energy.gov/wind/wind_potential.html (United States wind resource map produced by National Renewable Energy Laboratory) (last visited June 13, 2012); http://powerroots.com/pr/renewables/solar_photo-voltaic.cfm (United States photovoltaic resources map produced by the National Renewable Energy Laboratory) (last visited June 13, 2012); http://re.jrc.ec.europa.eu/pvgis/countries/countries-europe.htm (Links to maps of solar photovoltaic resource potential in Europe produced by the European Commission Joint Research Centre) (last visited Feb. 24, 2014).

2 The concept of externalities is among the most fundamental in basic micro-economic theory and is covered in introductory economics courses. *See, e.g.,* James R. Kearl, PRINCIPLES OF ECONOMICS 412–428 (D.C. Heath and Company, 1993).

3 For simplicity, this example focuses on an onshore commercial-scale wind energy project, although very similar analysis would follow for concentrating solar projects or even for smaller forms of renewable energy development.

4 To learn more about renewable portfolio standards, visit the United States National Renewable Energy Laboratory's webpage on this topic, available at www.nrel.gov/tech_deployment/state_local_activities/basics_portfolio_standards.html (last visited June 5, 2013).

5 The United States' Production Tax Credit (PTC) program was first introduced in 1992 and has been amended numerous times. Among other things, it has offered per-kilowatt-hour tax credit for energy generated from certain renewable sources, including wind. For a primer on the PTC and launching point for greater research on this topic, visit the U.S. Department of Energy's online Database of State Incentives for Renewables and Efficiency at http://dsireusa.org/incentives/incentive.cfm?Incentive_Code=US13F (last visited June 6, 2013).

6 For a detailed introduction to feed-in tariffs in Europe, *see generally* Miguel Mendonca, et al., POWERING THE GREEN ECONOMY (Earthscan, 2010).

7 For example, according to a 2012 Yale University poll, 78 percent of Americans believed that the United States should increase its use of renewable energy resources like wind, solar, and geothermal power. *See* Anthony Leiserowitz, et al., *Public Support for Climate and Energy Policies in September 2012* at 4, Yale Project on Climate Change Communication and George Mason University Center for Climate Change Communication (September 2012) (available at http://environment.yale.edu/climate-communication/article/Policy-Support-September-2012/ (last visited June 5, 2013).

8 For the author's own analysis of this topic, *see generally* Troy A. Rule, *Wind Rights Under Property Law: Answers Still Blowing in the Wind*, PROBATE & PROPERTY, Vol. 26, No. 6, pp. 56–59 (November/December 2012).

9 The author has also explored the topic of "solar rights" statutes in some detail. *See generally* Troy A. Rule, *Shadows on the Cathedral: Solar Access Laws in a Different Light*, 2010 U. ILL. L. REV. 851, 876–88 (2010).

10 The topic of wind turbine wake interference is addressed in far greater detail in Chapter 3.

11 To ease reading, footnotes from this excerpt were intentionally omitted.

2 Wind farms vs. neighbors

The worldwide expansion of the commercial wind energy industry over the past several years has been astonishing by any measure. From 2001 to 2011, the total installed wind energy generating capacity across the globe increased nearly tenfold, from 23.9 GW to more than 238 GW.[1] More installed wind energy generating capacity was added across the planet in 2011 alone than was even in existence just eight years earlier.[2] In this dawning golden age of wind power, images of bright turbines towering gracefully above green pastures have become iconic symbols of sustainability and modernity.

Humankind's growing infatuation with wind energy is also powerfully reflected in public opinion polls showing overwhelming positive support for wind power. According to a 2011 Eurobarometer survey, 89 percent of European Union citizens have a "very positive" view of wind energy.[3] A poll conducted in the United States in the same year found that 70 percent of Americans would favor additional wind energy development in their own communities—a greater level of support than for natural gas drilling, oil drilling, nuclear plants, or any other form of utility-scale energy development.[4]

The widespread appeal of wind energy comes as no major surprise given the tremendous benefits it can provide, particularly at the national and global levels. Once installed, wind turbines generate no new greenhouse gas emissions.[5] Wind energy development can also be a source of jobs, local investment, and economic growth and can allow nations to become less dependent on fossil fuels and foreign energy sources. Unlike nuclear power plants, wind farms produce little or no hazardous waste and do not make regions vulnerable to catastrophic events like the 2011 nuclear disaster in Fukishima, Japan, or the 2010 *Deepwater Horizon* oil spill in the Gulf of Mexico. And wind power is increasingly cost-competitive with more conventional energy sources, particularly in regions with strong wind resources and ample transmission capacity.

Unfortunately, although wind energy projects enjoy broad appeal in the abstract, they often garner far less favor within the communities where they are proposed. A 2010 study found that 30 percent of proposed wind projects in Europe were "stopped due to lawsuits and public resistance."[6] In many cases, the chief opponents of wind farms are individuals and companies that own or use land near the development site. Some of this local resistance is surely attributable to a generalized fear of the unknown—a generic "NIMBY" or "Not in My Back Yard" mentality that interferes with nearly every type of development at one time or another. However, in many instances the concerns of neighbors in the context of wind farm development are valid or at least worthy of being heard. As described in Chapter 1, many of the benefits of wind energy development accrue broadly to humankind throughout the world, while the social costs of wind farms often fall disproportionately upon those living next door to these massive projects. At least some of the local resistance that wind projects face seeks to compel developers and policymakers to consider and account for this inequitable distribution of benefits and costs during the development process. Experienced wind energy developers are well aware that the concerns of local residents must be heard, managed, and appropriately addressed for a project to succeed.

What are the greatest sources of local opposition to wind farms? And how can developers and policymakers address these sources of resistance and enable the continued growth of wind energy while still showing due regard for the rights of those living nearby? This chapter examines a laundry list of common conflicts between utility-scale wind energy projects and the activities of those living and working in their vicinities. Wind energy developers who educate themselves about these conflicts and learn to anticipate them in the early stages of a project can potentially avoid significant obstacles as the project proceeds. Policymakers who become familiar with these issues and enact statutes or ordinances to effectively govern them can reduce legal uncertainty and other unnecessary barriers to wind energy development within their jurisdictions.

Wind farms as economic dividers

Wind farms bring new money into rural areas, and this money is rarely distributed evenly among the local populace. This potential for a large infusion of money is generally viewed as a wonderful thing, but it can also be a source of tension and ultimately an obstacle to a proposed wind farm's success.

A newly proposed wind energy project can be particularly threatening to some rural dwellers because of its potential to create a new dynamic of "haves" and "have-nots" within their tight-knit community. Utility-scale wind farms are most frequently sited off of the beaten path and away from the big city. Because large-scale developments of any kind rarely occur in such rural areas, news of a potential wind farm tends to spread quickly among locals. Typically, some of these local residents will favor the project and some will not.

Landowners whose properties are leased for wind energy development can often experience large increases in their income if the project is built, while the income of landowners outside of the project stays largely the same. Although it is seldom explicitly discussed, this economic reality of wind farms can itself spark opposition to a new wind energy project. The most successful wind energy developers understand the social and cultural forces at play within host communities, and are able to anticipate how those differences might impact a community's overall willingness to accommodate a wind farm.

For most residents, the single biggest factor affecting their opinion of a proposed wind farm in their vicinity is whether their own land will be included in the project. Private landowners who expect to earn substantial new income under wind leases or easements if a project goes forward are often its most ardent supporters at the local level. Because a single commercial wind turbine can supply power to as many as 500 homes, enormous sums of money can be at stake for owners of these lands where turbines are proposed. For example, a Texas cattle rancher in the United States reported in 2008 that he received $500 per month for each of the 78 wind turbines on his property and anticipated that 76 more would be installed in the coming years.[7] Assuming that these other 76 turbines were added, this one rancher is earning $924,000 annually from wind energy development on his property.

Although the prospect of such handsome payouts is usually welcomed among those who would actually receive the money, it can turn excluded neighbors green with envy. Consider the plight of a hypothetical husband and wife whose land is situated just outside a proposed wind farm's site boundaries and is thus not being leased for the project. Suppose that this couple has long been close friends with several neighboring families whose lands *will* be leased for wind development if the developer's proposal succeeds. Historically, all of these households made roughly equivalent incomes by farming their respective acreages. As friends of similar socioeconomic status, they shared much in common and deeply enjoyed each other's company.

However, if the proposed wind project is developed as planned, that could all change. The couple's income will remain largely the same, while that of the neighboring families will dramatically increase. The market value of the couple's land might even decrease due to the project, compounding the seeming inequity of the situation.[8] One can only imagine the fears running through this couple's minds. Soon, their friends would be off spending their newfound wind lease income on fancy cars, expensive clothes, and extravagant vacations. Meanwhile, the couple would be left behind, staring out their windows at the rows of massive wind turbines towering across the street.[9]

Landowners like the fictional couple just described who own lands that are barely excluded from a proposed wind energy project are rarely fervent project advocates. Seeing few personal benefits from these projects and potentially viewing them as a threat to their relative social position and familiar way of life, some such landowners may even endeavor to hinder wind energy development for that reason. Recognizing that arguments about the general unfairness of their neighbors' new financial windfall are not likely to succeed, landowners like the couple are more likely to seek out other, more acceptable reasons to oppose wind farms in their communities.

As this book illustrates, a wide range of possible objections to wind energy projects exist, any one of which may be legitimate cause for limiting such development in certain contexts. However, when opposition allegedly based on aesthetic concerns, flicker effects, or other common complaints is actually driven by social or economic factors, developers and policymakers may be able to take steps to save a worthwhile project.

In any case, the first step is to determine whether a neighbor's particular objections to a proposed wind farm are valid. Although such inquiries are inherently fact-specific, the following sections provide basic background information on some of the most common neighbor-based complaints about wind energy. Many of these sections also mention some general strategies that developers and legislators may consider as they seek to address these problems.

Aesthetic objections to wind energy

Perhaps the most frequent objection that wind energy developers encounter is the concern that a wind farm could compromise the pristine visual beauty of an area. Numerous potential wind energy projects have been abandoned or significantly altered due to apprehensions over visual impacts, and some local governments have even cited viewshed protection as the primary reason for banning commercial wind energy development altogether in their jurisdictions. Whether a complete prohibition of wind farms based on view impacts is warranted in any particular locale is very difficult to assess, given that aesthetic preferences are inherently subjective and hard to measure. However, as evidenced by the ever-evolving nature of fashion trends, individuals' opinions about what is beautiful and what is ugly can change over relatively short periods of time. For commercial wind energy development to continue its rapid growth rate, it will eventually need to move closer to population centers where aesthetic issues will only become more pronounced. Effectively managing these concerns and persuading more of the world to accept the presence of these massive structures in the skyline will be ever more crucial to the long-term growth of wind as an energy source in the next century.

Without question, the gargantuan wind turbines installed in today's commercial wind farms can materially alter a landscape's appearance. Modern utility-scale wind turbines commonly exceed 400 feet in height, towering well above any other buildings or structures in their vicinities and tall enough to be seen from several miles away.[10] Even in rural areas where population densities are relatively low, turbines can impose significant costs by disrupting territorial views for local residents who may have grown attached to an area's existing natural backdrop. The presence of turbines continues into the night, when turbine safety lighting often required under federal aviation laws flashes across an otherwise pristine evening sky.[11]

Unfortunately, only so much can be done to disguise commercial wind turbines from view. Because the colors naturally occurring in the sky and on land tend to change with the seasons and time of day, it is often impossible to successfully camouflage turbines with paint such that they blend in with their surroundings.[12] Painting designs on turbines or painting them multiple colors tends to only make them more distracting, and painting them gray can make them seem "dirty" or "associated with an industrial, urban or military character."[13] Consequently, most commercial wind turbines are painted white—a color choice based partly on a belief that bright white turbines "convey a positive image" and are "associated with cleanliness."[14]

Installing smaller, shorter turbines to make them less conspicuous to neighbors is also rarely a viable option. The energy productivity of natural wind tends to increase significantly with altitude, so turbines are purposely designed to stand high above the ground to capture those more productive wind currents.[15] By towering well above the earth's surface, modern commercial wind turbines also avoid turbulence from nearby buildings and trees that might otherwise diminish their productivity.[16] And the sheer size of a commercial wind turbine's rotor, which directly affects its generating capacity, requires that the turbine be mounted upon a tall tower.

Unable to camouflage or shrink the size of utility-scale wind turbines, wind energy developers must often find ways to assuage locals' concerns about the potential visual impacts of these enormous devices. Developers' ability to do so depends in part on local residents' subjective views about the attractiveness of the turbines themselves. Indeed, wind turbines are no different from any other structure in that their beauty or ugliness ultimately rests in the eye of their beholder. Some scholars have suggested that wind farms could and should be more commonly viewed as works of art.[17] Citing the widespread depiction of windmills in notable seventeenth-century Dutch paintings and the large-scale environmental art projects of famous artists such as Christo Javacheff, they argue that commercial wind energy projects should be perceived as artistic creations rather than industrial blight.[18]

Of course, not all landowners near wind farms show such appreciation for wind turbines in their communities. In some instances, neighbors who fear the visual impacts of wind energy projects have sought to stop them by filing aesthetics-based nuisance claims. Nuisance claims generally require a showing that the defendant's unreasonable conduct has interfered with the plaintiff's own reasonable use of land.[19]

Fortunately for wind energy developers, courts in many countries are reluctant to find a nuisance against a productive land use like a wind farm based solely on aesthetic objections. Courts in the United States have generally held that the mere fact that a "thing is unsightly or offends the aesthetic sense of a neighbor does not ordinarily make it a nuisance or afford ground for injunctive relief."[20] Still, at least one court in the United States has suggested that neighbors to a commercial wind farm could recover for nuisance based on the "turbines' unsightliness."[21] Wind energy developers in the United States thus cannot have total certainty that they are protected from aesthetics-based nuisance claims.

Well aware of the potential for aesthetic concerns to slow the pace of onshore and offshore wind energy development, the International Energy Agency (IEA) has begun searching for ways to change popular perceptions about the appearance of wind farms. In May of 2012, the IEA distributed a report summarizing a three-year study on how to encourage greater social acceptance of wind farms.[22] The working group that prepared the report consisted of representatives from 10 countries throughout the world who met semiannually to discuss this issue.[23]

Among other things, the IEA report concluded that some local opposition to wind farms is useful in that it can be a source of localized knowledge that can ultimately make a project more socially beneficial.[24] Still, the report described the need for greater "monitoring" of social acceptance of wind farms across cultures and projects.[25] As wind energy continues to spread across the world, more and more information will become available for formulating best practices to promote social acceptance of these projects in the coming years. In the meantime, the wind energy industry and other wind energy supporters must continue searching for ways to change popular perceptions about aesthetic impacts of wind turbines, characterizing them as "elegant and graceful ... symbols of a better, less polluted future."[26]

The Cape Wind project and aesthetics-based opposition to offshore wind energy

Historically, most commercial wind energy development has occurred onshore so aesthetic objections to such projects have typically related to terrestrial viewsheds. However, as activities surrounding the infamous "Cape Wind" project in the United States have shown, aesthetics-based opposition can hamper or halt wind energy development offshore as well.

Cape Wind is a proposed offshore wind energy project that would involve roughly 130 turbines spread out among 25 square miles near the center of the Nantucket Sound off the coast of Massachusetts in the United States.[27] This "Horseshoe Shoals" area of the Sound is characterized by shallow water depths, exceptional wind resources and a central location in the Sound relatively distant from sensitive habitat or fishing areas. Convinced that these factors made the area a seemingly ideal place for an offshore wind farm, developer Jim Gordon filed a permit application for a project there in 2001.[28] At the time, it seemed almost inevitable that Cape Wind would be the first commercial offshore wind farm in the United States, supplying about 75 percent of the power needs in and around Nantucket, Martha's Vineyard, and Cape Cod.[29]

The only major downside of the Cape Wind project is its proximity to several affluent residential areas. Anticipating aesthetic objections to the project from the residents of these communities, Cape Wind's developer proposed to paint the project's turbine towers to blend in with the surroundings and to install its turbines at least five miles away from shore so that they would appear no taller than a half inch from the shoreline.[30]

Unfortunately, the developer's proposed visual mitigation measures were not nearly enough to appease the majority of residents of this region, which has been described as the home to "some of the wealthiest and most powerful families in the United States."[31] Concerned primarily about Cape Wind's potential impacts on their beachfront views and on the value of their homes, opponents of the project employed a wide variety of strategies for more than a decade to slow its progress. Opponents challenged the jurisdiction of agencies that purportedly approved certain aspects of the project.[32] They disputed the FAA's determination that the project would not create hazards for aviation.[33] They even asserted that the project lacked a necessary state fishing permit.[34] Although construction on Cape Wind is scheduled to finally begin in 2014,[35] the developer has lost millions of dollars because of challenges that seem ultimately rooted in concerns about the project's perceived impacts on ocean views and local property values.

For obvious reasons, offshore wind energy projects are particularly vulnerable to aesthetics-based opposition. People throughout the world have long prized the uniquely untouched, calm nature of high-quality beach-front views. Many home buyers are willing to pay extra for homes that have views, too—nearly 60 percent more than for homes without ocean views, according to one popular study.[36] It is thus hardly surprising that owners of beachfront homes are less than enthusiastic when a new offshore wind farm threatens to degrade these views, even if the degradation is slight.

In an attempt to address conflicts over the aesthetic impacts of offshore wind energy, the governments of Germany and The Netherlands have enacted laws prohibiting the siting of offshore wind farms within 12 miles of the shore.[37] Even on the clearest days, turbines sited that far out at sea are likely to be undetectable with human eyes from the beach. The laws of these countries avoid conflicts between offshore wind energy and views, but they do so at a hefty price, effectively preventing the development of vast amounts of valuable wind resources.

There is reason to believe that there are better ways of dealing with these conflicts than to simply prohibit any offshore wind development across huge stretches of coastal waters. More efficient policies in this context seek to balance the competing interests of beach view protection and offshore wind development through marine spatial planning and related programs. These programs, which are comparable to zoning laws for land, use maps to coordinate various competing uses of marine areas within a large body of water. The EU created a roadmap for greater spatial planning in Europe's water bodies in 2008, and some countries in Europe have already adopted marine spatial plans to help promote more fair and efficient use of sea resources.[38] Such planning will grow ever more important as a tool for managing potential conflicts with beachfront views, shipping routes, fishing operations, habitat preservation, offshore oil and gas development, and other marine uses as offshore wind energy development continues to spread in the coming decades.

Wind turbine noise and health impacts

Concerns over turbine noise are another common source of opposition to wind farms. Commercial wind turbines undeniably generate some amount of noise. A turbine's mechanical noise, which typically originates from its gearbox and other mechanical parts, is not typically the main source of noise complaints at a wind farm.[39] More often, these complaints relate to the aerodynamic noise a turbine can generate when its blades rotate through the air.[40] The relative loudness of a wind turbine depends to a large extent on the size and design of the turbine itself. For any given turbine design, the noise level it produces can also vary based on prevailing wind speeds and the turbine's angle in relation to the wind.

Over the past several years, academicians across the planet have conducted dozens of studies on wind turbine noise and its possible effects on humans and animals. This chapter does not attempt to summarize the voluminous amount of published scientific and empirical research on this topic, which would go well beyond the scope of this book. Instead, what follows is a brief primer on the main issues surrounding the wind turbine noise debate, which should be sufficient to educate readers on the basics and to provide a launching point for those interested in further investigating this important subject.

Just how noisy is a wind farm?

Wind energy development can introduce noise into a community in two primary ways. First, the wind farm construction process involves large machinery and activities that inevitably produce some noise. For example, some studies suggest that noise associated with "pile driving"—the driving of support poles into the seabed floor—is the "most significant environmental impact" associated with offshore wind energy development.[41] Temporary construction noise is obviously not unique to wind energy—it can occur in connection with almost any type of major construction project. Nonetheless, such noise can substantially interfere with neighboring land uses and thus often warrants consideration during the development process. Commonsense measures requiring developers to perform the loudest construction work during waking hours and to provide advance warning to landowners who are most likely to be affected can help to prevent temporary construction noise from unnecessarily slowing progress on a wind energy project.

The other primary source of sound disturbance from wind energy development is the audible noise that turbines produce while in operation. Sound is generally described through two primary measures: loudness and pitch. The loudness of sound is typically measured in decibels (dB), while the pitch or frequency of sound is measured in Hertz (Hz).[42] Audible turbine noise is often measured in "A-weighted" decibels (dBA), which "de-emphasiz[e] the very low and very high frequency components of the sound in a manner similar to the frequency response of the human ear."[43] Sounds registering only 10 dBA are barely audible to human ears, and sounds exceeding 130 dBA are noticeably painful.[44]

At a distance of 300 meters from the base of a commercial wind turbine—a typical minimal setback distance for utility-scale wind turbines from businesses or homes—turbines generate a noise level of only about 43 decibels.[45] That level of noise is only slightly louder than the sound of a running refrigerator and is quieter than the sound of a mid-size window air-conditioning unit.

Although wind turbine noise is generally modest in comparison to many other common sources of ambient noise, it can be a source of significant disturbance in some contexts. Unlike construction noise, which is inherently intermittent and may only last for a few weeks or months, turbine noise is incessant. The setting where the noise occurs can also be an important factor, as sounds of any type are more likely to draw opposition in the sorts of rural communities where wind farms tend to be sited. Many urban dwellers have become desensitized to ambient noise levels of 40 decibels or more because such levels are common in bustling cities and are detectible at almost any hour of the day. In contrast, rural landowners are often more accustomed to quiet surroundings and may thus be more apt to take issue with the low hum or pulse of wind turbines.

Disputes over turbine noise

As commercial wind energy development has rapidly expanded over the past decade, disputes over wind turbine noise have begun surfacing across the globe.[46] In 2010, a French court ordered that eight commercial wind turbines be shut down during sleeping hours to address noise concerns.[47] Hundreds of complaints about wind turbine noise have been made in Great Britain in recent years, and some such complaints have led wind farm operators to shut down turbines whenever winds reached a certain speed.[48] And noise-based wind farm opposition has been a persistent obstacle for developers in the United States, where landowners have bemoaned wind turbine noise in several different states.[49]

Unlike land uses that are offensive based solely on their aesthetic attributes, noises that unreasonably disturb neighbors are widely recognized ground for common law nuisance claims in the United States.[50] In the famous case of *Rose v. Chaikin* in 1982, a court in the state of New Jersey enjoined operation of a 60-foot tall wind turbine that had been installed just ten feet from the plaintiff's own property line.[51] The plaintiff complained that noise from the turbine was causing "stress-related symptoms" including "nervousness, dizziness, loss of sleep, and fatigue."[52] The court expressly noted the social benefits of wind energy and that such benefits must be weighed in such situations.[53] However, citing longstanding case law for the principle that noises did give rise to an actionable nuisance if they caused unreasonable "injury to the health and comfort of ordinary people in the vicinity," the court ultimately held in favor of the plaintiff and prevented further use of the turbine.[54]

Although *Rose v. Chaikin* involved a small turbine, courts in the United States have recognized that noisy commercial wind farms can constitute a common law nuisance as well. In *Burch v. NedPower Mount Storm, LLC*, the West Virginia Supreme Court overturned a lower court's dismissal of a nuisance claim based on the noise of a large, utility-scale wind energy project near the plaintiffs' land.[55] According to the court, a plaintiff may be entitled to an injunction under West Virginia State law for any noise that "prevents sleep or otherwise disturbs materially the rest and comfort" of nearby residents.[56] Accordingly, developers in the United States must consider the possibility of a nuisance claim if their turbines generate that level of noise at a residence nearby.

Low-frequency "infrasound" and the wind turbine syndrome debate

In addition to ordinary noise, wind turbines can also produce less perceptible noise that some believe could be equally detrimental to neighbors of commercial wind farms. In recent years, a handful of researchers have suggested that certain *inaudible* sound waves emanating from wind turbines might be injurious to humans. Infrasound, which occurs at frequencies less than 20 Hz, is not perceptible by most people at the decibel levels produced by commercial wind turbines. However, operating turbines can produce such low-frequency sound, and this infrasound has been the subject of much debate in recent years.

In 2009, Dr. Nina Pierpont published a book on what she labeled "Wind Turbine Syndrome", creating some amount of stir within the wind energy industry.[57] The book summarized the author's research concluding that the infrasound waves generated by rotating commercial wind turbines were capable of causing a wide range of symptoms such as migraine headaches, vertigo, motion sickness, gastrointestinal sensitivity, and anxiety.[58] Dr. Pierpont has since conceded that the low-frequency noise she describes in her book is not uniquely produced by wind turbines and can be created by natural gas compressor stations, industrial sewage pumping stations, industrial air conditioners, and other land uses.[59] However, Pierpont continues to stand by her claim that these inaudible sound waves are the root cause of various symptoms reported by some people living near wind energy developments.

The publication of Pierpont's book in 2009 and the growing dialogue about the potential adverse health impacts of wind turbines at that time posed a significant threat to the wind energy industry, playing directly to the fears of landowners about possible unknown effects from this relatively new type of land use. In response to these developments, two large wind energy industry groups commissioned their own study on the topic. In December of 2009, the American Wind Energy Association (AWEA) and Canadian Wind Energy Association (CanWEA) distributed a report entitled *Wind Turbine Sound and Health Effects: An Expert Panel Review.*[60] Not surprisingly, the lengthy report compiled by a group of seven Ph.D.'s and medical professionals from around the world reached conclusions that contradicted those of Pierpont and her followers. Among other things, the report found "no evidence that the audible or sub-audible sounds emitted by wind turbines have any direct adverse physiological effects" and that "ground-borne vibrations from wind turbines are too weak to be detected by, or to affect, humans."[61]

The AWEA/CanWEA report summarized volumes of peer-reviewed research that directly contradicted Pierpont's assertions regarding the myriad alleged health effects of the low-frequency sound waves detected near wind farms. In all, the report has proven fairly effective at assuaging fears about alleged Wind Turbine Syndrome. Still, the wind industry's responsive study is afflicted by an unavoidable appearance of bias because it was not truly independently produced. Already, numerous additional studies have been conducted that seek to further resolve questions about the possible health effects of low-frequency turbine noise.[62] As more independent research on this issue follows in the coming years, the world will have greater certainty regarding whether the low-frequency waves emanating from wind turbines actually pose any harm to those living nearby.

Shadow flicker effects

Even landowners who are too far away from commercial wind turbines to be bothered by noise can still be irritated by the long, moving shadows of wind turbine blades. At certain times of day and seasons of the year, rotating wind turbines can cast revolving shadows for up to two kilometers that create a strobe-like effect through the windows of homes and other buildings.[63] These "flicker" effects from commercial turbines seldom persist at the same location for more than a half hour on any given day and do not typically pose a serious danger or health threat to humans or animals. Still, they can unquestionably disturb or annoy those living or working close to wind farms and must be considered during the development and approval process.

Flicker effects occur when direct sunlight is blocked by the blades of a rotating turbine, creating shadows that move as the turbine blades move. As sunlight flashes repeatedly between the revolving blades, a pulsing light-to-dark-to-light effect results. Whenever the sunlight passing through windows is a room's primary light source, this flicker effect can create the appearance of lights recurrently being turned on and off.

Because most commercial turbines are sited a significant distance from buildings, flicker effects tend to cause disturbances primarily when the sun is low in the sky. A low sun most commonly occurs at sunrise and sunset and is more frequent during winter months and in regions located far from the equator.[64] Clouds and overcast conditions can prevent or substantially mitigate flicker effects, but on clear days these effects can seriously disrupt neighbors' quiet enjoyment of their land.

Aside from causing general annoyance, wind turbine flicker effects can theoretically create serious medical risks for the small percentage of the global population that suffers from photosensitive epilepsy.[65] Those with this condition can suffer epileptic seizures when exposed to flickering sunlight or other rapid flashes of light. Fortunately, most researchers have concluded that the utility-scale wind turbines installed on wind farms today are so large that their blades do not ordinarily rotate at rates fast enough to trigger epileptic seizures.[66] Despite the existence of tens of thousands of wind turbines across the planet, as of mid-2012 there had not been any widely reported instances of seizures triggered by turbine shadow flicker.[67]

In cases where a nearby wind farm's flicker effects are already substantially disrupting a landowner's enjoyment of land, some landowners are likely to seek a legal remedy. In the United States, such situations raise an interesting legal question because the U.S. common law has long held that landowners are not liable for damages caused by their casting of shadows onto a neighbor's land. In the famous U.S. case of *Fountainebleau Hotel Corp. v. Forty-Five Twenty-Five, Inc.*, the Eden Roc Hotel in Miami Beach, Florida, sued a neighboring hotel owner to enjoin its construction of a 14-storey addition that would cast a large shadow on the Eden Roc Hotel's beachfront property. The court refused to recognize a nuisance claim based on the new shading, declaring that the Eden Roc had no legal right to the free flow of light from adjoining land.[68] Based on this longstanding precedent, landowners in the U.S. may have difficulty succeeding on common law nuisance claims based on flicker effects, which are essentially disruptions of the free flow of sunlight onto land.[69]

On the other hand, the shadows at issue in flicker disputes and the nature of the alleged resulting nuisance are fundamentally different than those in *Fountainebleau*. The wind turbine shadows that cause flicker effects are constantly moving and can create an annoying distraction in and of themselves, which is arguably more worthy of recognition as a nuisance than mere unwanted shade. At least one U.S. court has seemingly recognized this difference, overturning a lower court's dismissal of a nuisance claim that was based in part on a wind farm's flicker effects.[70]

Fortunately, computer programs now exist that are capable of accurately predicting flicker effects at prospective turbine sites. One commonly-used software program is WindPRO, a platform developed by Denmark-based EMD. WindPRO uses information on the proposed types and locations of a project's proposed turbines, the surrounding topography, and other factors to predict the precise dates and durations of potential shadow flicker events.[71] Such information can be a powerful tool for identifying potential problem sites during the early stages of development so that turbine sites can be adjusted accordingly. Reports generated through WindPRO and similar software programs can also be instrumental in allaying neighbors' concerns about shadow flicker during the turbine siting and permitting process.

Given the ever-improving quality of shadow flicker software and the availability of reasonable strategies for mitigating flicker effects, shadow flicker disruptions are among the most easily resolvable of all potential land use conflicts between wind farms and neighbors. Armed with these tools, developers can generally prevent these disturbances from arising in the first place. In cases where adjusting proposed turbine sites based on flicker effects is not economically justified, wind farm operators may also be able to program turbines to shut off during those relatively short times of day during periods of the year when shadow flicker problems are possible. Planting new trees or vegetation can also shield and further protect buildings that may be vulnerable to flicker problems.

Many jurisdictions have already enacted laws that adequately protect against shadow flicker conflicts at wind farms. In some jurisdictions, these laws simply limit the total number of allowable hours of turbine shadow flicker at a given location per year. Laws in other locales completely prohibit shadow flicker on residential homes.[72]

As policymakers become better educated about shadow flicker and take steps to address it in their jurisdictions, these conflicts should become less and less common over time. A 2011 report on independently commissioned research on shadow flicker for the United Kingdom's Department of Energy and Climate Change (DECC) reached similar conclusions. Researchers found that shadow flicker was not a significant problem for wind farms in the UK, does not pose any major health risks, is easily mitigated through turbine shut down systems, and is already adequately addressed in the DECC's existing planning guidance on the topic.[73]

Ice shedding and ice throws

Another concern that neighbors commonly raise in the context of wind energy development is a fear that turbines installed on nearby land could throw snow or ice across the property line, causing injury or property damage. Like skyscrapers, utility poles, and most other large structures, wind turbines can accumulate snow and ice on them on during wet, chilly days. In some instances, rotating wind turbine blades have thrown large chunks of this ice hundreds of meters from turbine towers.[74] High winds can also strip ice from a turbine blade and carry it a significant distance from the turbine on its descent to the ground.[75] In colder climates or high-elevation areas, this risk of thrown or fallen ice can be yet another reason for locals to oppose a proposed wind energy project.

Several researchers have studied the ice-related risks of wind turbines over the past decade. In 2007, a group of researchers disseminated an article describing their mapping of ice throws and falls below a 600 kW wind turbine near a ski resort in the Swiss Alps.[76] The researchers inspected the ground below the turbine after every "icing event" for two consecutive winters, documenting the weight, size and location of every chunk of fallen ice. They found 94 ice fragments over the course of just one winter during the test period, and some of the fragments they found were as long as 100 centimeters and weighed as much as 1.8 kilograms.

Siting turbines an adequate distance away from buildings and residential areas is the most basic way to prevent damage or injury from ice throws. Although wind speed and the shape of the ice fragment can also affect the distance that ice travels when thrown from a turbine blade,[77] the following general formula has been offered by some as a means of calculating minimum setback distances for protection from ice throws:[78]

Minimum setback distance = 1.5 * (hub height + rotor diameter)

Although utility-scale wind turbines are often set back at least this distance from residential buildings, they are sometimes sited closer to roads and other public areas where people could be put in danger. Additional setback requirements from these other land uses may be warranted in climates and settings where turbine blade ice is likely to be a problem. For example, in the case of the wind turbine studied in the Swiss Alps, operators placed additional restrictions on public access to the area near the turbine as a precaution against injury to the public.

It is worth noting that turbine ice throws could theoretically pose a threat to livestock grazing near wind turbines as well. Since only a very small percentage of the total surface area involved in a wind energy project is physically occupied by turbine towers or supporting infrastructure, ranchers often continue to allow cattle or other animals to graze on land that has been developed for wind energy. Although this practice allows for more overall productivity of range land, it can result in cattle grazing directly under the enormous rotating blades of commercial wind turbines where ice could fall and cause injury. Fortunately, risks associated with ice throws are often allocated between developers and landowners in wind energy leases and existing setback restrictions in most areas make the risks to neighboring landowners' cattle relatively low.

Increasingly, modern commercial wind turbines are equipped with sensors that can detect vibrations or the additional weight of ice on their blades and are programmed to stop operating when blades get icy.[79] However, anecdotal evidence suggests that these sensors don't always shut off turbines in time to prevent ice throws, putting nearby humans and property in danger.[80]

In addition to following adequate setback requirements and equipping turbines with sensors that stop turbines when their blades accumulate ice, there are a few other ways to potentially mitigate risks associated with falling ice and ice throws. Fencing off areas immediately below turbines, prominently displaying warning signs near wind farms, and equipping visitors with protective head gear are relatively low-cost means of providing added protection against ice-related injuries.[81]

Where appropriate, painting turbine blades black may even help to reduce ice buildup and marginally reduce the risk of thrown ice.[82] The color black absorbs more sunlight than white or gray, so black turbine blades tend to stay somewhat warmer and are slightly less apt to collect ice. In cold regions or higher elevations where icing is a major problem for much of the year, black blades might thus be worthwhile as a means of reducing the frequency of de-icing and its accompanying costs. Of course, painting turbine blades black can also make them more conspicuous in the landscape, which may often outweigh the ice-related benefits of that approach.

Interference with military radar

Sometimes, the "neighbors" disrupted by wind farms do not reside literally next door but are military bases or weather monitoring stations situated dozens of miles away. These distant parties can sometimes object to new wind energy projects because of their potential to disrupt the transmission of electromagnetic waves through low-altitude airspace.

Radar systems and other modern electronic devices are capable of trans-mitting electromagnetic waves of varying frequencies through air.[83] When a massive wind turbine is erected in a given area, it can block or alter electro-magnetic waves at some frequencies, interfering with their transmission. Wind turbines have the potential to disrupt microwave beam communi-cation systems and can impair reception for certain types of television, radio, and cell phone signals.[84] They can also confuse meteorological radar and complicate weather forecasting.[85] However, the most contentious clashes between wind farms and radio wave transmission have involved military radar systems, which use radio waves to detect the location, size, and speed of distant objects.[86]

Regulations aimed at avoiding conflicts between commercial wind turbines and military radar have significantly hindered wind energy development in recent years, especially in the United States. Many of the most promising sites for wind energy projects in the United States are located in western regions of the country where wind resources are strong and land is plentiful. Unfortunately, these wide open spaces are also prime locations for military training bases and other military facilities that make heavy use of radar equipment. As wind energy development has rapidly expanded in the western United States over the past decade, it has often encountered resistance from military interests concerned about the potential impacts of wind farms on their radar systems.[87]

Under regulations issued by the United States Federal Aviation Administration (FAA), the U.S. Department of Defense (DOD) can formally challenge proposed commercial wind turbine installations that they believe could interfere with military radar.[88] These regulations require parties to file a Notice of Proposed Construction with the FAA before erecting any structure that extends more than 200 feet above the surface of land.[89] The FAA then examines the site and decides whether to issue a Notice of Presumed Hazard (NPH) based on its determination that the proposed structure could create a hazard within navigable airspace.[90]

If the FAA issues an NPH against a wind energy developer, a potentially long and expensive process can ensue, potentially delaying progress on wind farm projects. In some instances, the FAA has issued an NPH against a wind energy project even when the proposed turbines would be sited dozens of miles away from any military base.[91] AWEA claims that roughly half of the generating capacity of all commercial wind farms proposed in the United States in 2009 was "abandoned or delayed because of radar concerns raised by the military and the [FAA]."[92]

One of the most highly-publicized clashes between a wind energy developer and U.S. military interests involved the $2 billion Shepherds Flat project in the U.S. state of Oregon in 2010. When the Shepherds Flat project was announced, it promised to be the largest wind farm in the country, involving more than 300 commercial-scale turbines.[93] Less than two months before its developer planned to break ground on the project, the FAA issued an NPH that prohibited the erection of any turbines in the vast project area.

The FAA's objection to the Shepherds Flat project was based on concerns from the U.S. Air Force that the project could interfere with its antiquated radar station in Fossil, Oregon, about 50 miles away.[94] Delays associated with the FAA's NPH threatened to kill the enormous project, which had already been nine years in the making.[95] After senators from the state of Oregon and members of President Barack Obama's cabinet put significant pressure on the DOD, military officials ultimately retracted their objections to the Shepherds Flat project, enabling it to proceed.[96]

In many cases, military interests can easily avoid conflicts between their radar and private wind energy development by installing relatively inexpensive upgrades to aging radar systems.[97] With the help of researchers, the DOD ultimately concluded that such upgrades could prevent radar interference problems near the Shepherds Flat project. The UK has also been upgrading its radar systems to prevent interference with wind energy development.[98] Studies have shown that the cost of such radar upgrades is often far less than a wind farm's potential benefits.[99] "Stealth" turbines designed to minimize interference with radar systems will also be hitting the market in the coming years and could further mitigate disputes between military interests and wind farm developers.[100]

Although the dispute at Shepherds Flat was resolved and technological fixes to the radar disruption problem are often available, these conflicts could continue to create obstacles for developers in areas of the world that have yet to update their radar systems. In mid-2012, U.S. Navy officials indicated that they would oppose any wind energy project within a vast proposed "adverse impact zone" area covering much of California's Mojave Desert and portions of Nevada, based on radar interference concerns.[101]

Interestingly, local governments might occasionally oppose wind farms to avoid possible radar interference conflicts that could endanger a nearby military base that is important to the local economy. At least one municipal government in the U.S. has allegedly denied approval for a wind project for the primary purpose of protecting a nearby military base from closure. Such local officials supposedly believed that potential radar interference problems from the proposed wind project could make the military base more vulnerable during the federal government's next round of base closures.[102]

In the United States, Congress enacted new provisions in the 2011 National Defense Authorization Act aimed at addressing some of the issues associated with conflicts between wind farms and military radar. Among other things, the legislation required the DOD to create a clearing house for wind energy siting approvals. Under these provisions, developers are no longer required to contact multiple military bases about their projects, and the DOD must make determinations on applications within 30 days.[103] Such changes should enable developers to avoid the sort of last-minute military resistance that put the Shepherds Flat project in jeopardy.

Interference with mineral rights

In addition to disrupting users of neighboring land and airspace, wind farms can sometimes interfere with property rights in assets located immediately below the project itself. When known deposits of oil, gas, or other valuable minerals exist under parcels that are targeted for a wind energy project, conflicts can arise over how to accommodate both mineral extraction and wind energy on the same land.

Unlike laws in most other countries of the world, United States law provides that landowners hold private property rights to extract and sell minerals below their land. Particularly in western portions of the United

States, many landowners have severed these "mineral rights" from their surface rights and conveyed them to other parties. Such mineral rights generally include limited rights to access and disturb the surface of the land for purposes of exploration and extraction.

Developers of utility-scale wind energy projects in the United States sometimes discover during early stages of the planning and development process that parties other than the wind energy lessor have interests in oil, gas, or minerals deposited below the leased land. These subsurface interests are usually disclosed as special exceptions on commitments for title insurance. In many states, these mineral rights are "dominant" over the surface estate, meaning that they take reasonable priority over the surface owner's rights. They create additional risks for wind energy projects because mineral rights holders or oil and gas lessees could theoretically seek to exercise their exploration or extraction rights in ways that interfered with the wind farm or its development. Because lenders and investors are unlikely to finance a project unless risks associated with oil, gas, or mineral rights are addressed, developers and their legal counsel must find ways to deal with them. Title insurance companies are often willing to provide insurance against oil- or mineral-rights related risks, but only if such companies are convinced that the risks are minimal.

One recent case may provide wind farm developers in the United States a small amount of additional comfort as it relates to mineral rights on their project lands. The plaintiff in *Osage Nation v. Wind Capital Group, LLC*, was a Native American tribe that held mineral rights on certain lands that the defendant had leased for wind energy development.[104] The tribe sought to enjoin development of the wind farm, asserting that its mineral rights were the dominant estate—superior to the surface rights associated with the defendant's wind farm lease—and that the wind farm would limit the tribe's ability to develop its mineral estate. The court rejected the tribe's arguments, concluding that the tribe failed to establish that the wind farm would unreasonably interfere with any exercise of mineral rights on the property.[105] The wind farm would allegedly occupy less than 1.5 percent of the surface within the project area, and oil drilling technologies were available to avoid any potential conflict with wind turbines.[106] Based on these sorts of arguments, the court concluded that minerals extraction and wind energy development could harmoniously subsist on the same parcels of land.

In some instances, developers or their legal counsel may be able to contact mineral rights holders and convince them to sign an agreement promising not to disturb the wind farm. Depending on the context, mineral rights holders may conclude that it is in their interest to sign such non-interference agreements. Often, drilling and exploration for oil and gas occurs at a parcel's lower elevations, while wind turbines tend to be installed at higher elevations where more productive wind generally blows. On lands where these generalizations hold true, mineral rights holders may be less concerned about potential conflicts with wind farms. In fact, mineral rights holders may actually benefit from the wind project because it would result in new roads and other infrastructure that may lower the costs of exploration and extraction on the land. Mineral rights holders may also feel pressure from surface rights holders to cooperate and sign non-interference agreements because surface owners often stand to make lots of money if a wind project proceeds. If a mineral rights holder unjustifiably blocks the wind project, the owner of the surface estate may prove less eager to accommodate future mineral exploration or extraction activities on the land.

Severed wind rights?

Conflicts between wind energy developers and mineral rights holders are likely to become even more challenging in situations when a surface owner has purportedly severed both the mineral rights and the "wind rights" from the surface estate. As wind energy development has spread throughout the western United States over the past decade, numerous landowners have claimed to sever wind rights from their surface estate. When this has occurred, a developer can find itself attempting to balance the interests of *three* different parties: the wind rights holder, the surface rights holder, and the mineral rights holder. Surface owners in these situations can find themselves stuck in the middle when minerals extraction and wind farms development are both occurring on their land.[107] If they have nothing to gain from successful wind farm development or mineral rights extraction on their property, these surface owners may actually prove reluctant to assist in resolving disputes between mineral rights holders and wind developers.

The practice of wind rights severance can also raise difficult new legal questions regarding whether the minerals estate or the wind estate for a given parcel is the "dominant" estate. In other words, when a wind rights holder wishes to install a turbine at a specific location but a mineral rights holder wants to extract oil or minerals from the same spot, who wins? Although courts in the United States seem to have reached some agreement as to the priority relationship between mineral estates and surface estates, introducing a wind estate into the mix creates an entirely new set of priority issues.[108]

Perhaps in light of the many perplexing issues wind severance can raise, a growing number of state legislatures in the U.S. have recently enacted laws explicitly rejecting the concept of wind rights severance. In states that have enacted such legislation, efforts to balance the interests of wind energy developers and mineral rights holders will be a bit less complicated in the decades to come.[109]

Interference with crop dusting

Crop dusting is yet another activity that can be impacted by wind energy development. Some farmers periodically use small airplanes to aerially apply pesticides, fungicides, or fertilizers onto rural land through a process commonly known as "crop dusting." When a wind farm is added to a farming community, crop dusters must obviously be careful to avoid collisions with the new turbines. However, wind farms also create another, less visible risk that can be equally dangerous for crop dusting planes: wind turbines create a wake of turbulent airflow behind them that can be hazardous to small airplanes flying up to half a mile away.

Because of the additional risks associated with crop dusting near wind farms, some crop dusting companies in the United States have been known to charge as much as a 50 percent surcharge to crop dust near wind turbines.[110] At least one wind energy developer has offered to directly pay these surcharges in an effort to appease neighboring farmers.[111] Still, opposition from crop dusting companies and from neighboring farmers who rely on their services can create additional obstacles for wind projects sited near farmland.

Above-ground transmission lines commonly associated with wind farms can also disrupt crop dusting activities. Transmission lines are often buried below the ground on the project site but are then brought above ground once electricity passes through the project's substation and its voltage increases. Some of these lines inevitably must cross through agricultural fields under easements, and occasionally these lines get in the way of crop dusters. A crop dusting company in Oklahoma, United States, recently sued a wind energy developer because of concerns that a new 90-foot transmission line for the developer's wind energy project would inhibit the company's crop dusting flights.[112] The transmission line was to be erected relatively close to the company's private airstrip. Although the developer held transmission line easements for the line and the FAA had already approved it, the lawsuit created an unexpected obstacle to completing the project.

A few possible strategies have emerged for dealing with conflicts between crop dusting and wind farms, although none of them offer a panacea for this problem. Ground-based crop spraying may be one means of mitigating these conflicts, although many farmers prefer aerial spraying because it is less damaging to crops. One way of avoiding the hazards of turbine wakes would be for operators of wind energy projects to agree to shut off turbines during crop spraying upon reasonable notice, although even that would not eliminate the risk of collisions.

Given that wind energy projects are frequently developed in agricultural areas, it is not surprising that many wind energy leases have provisions that directly address the topic of aerial spraying in an effort to prevent it from later becoming an issue of contention. These tensions are likely to only continue as wind energy development moves increasingly beyond rangelands and into farmlands over the next century.

Interference with hunting

In rural communities where locals have been hunting in the same general areas for many years, a wind project's potential impacts on hunting can also be a significant source of resistance. Bird hunting is a popular sport on many rural lands. Hunting for deer, small mammals, and other wildlife is also not uncommon on remote rural property.

Introducing large wind turbines onto land can limit its availability for hunting for multiple reasons. Hunters themselves may face safety risks when they are in close proximity to turbines and their supporting facilities. Similarly, hunters' stray bullets can put wind farm employees at risk and can damage project facilities. To the extent that the presence of turbines drives wildlife out of an area, wind farms can also upset neighbors who feel that introduction of turbines nearby has degraded hunting opportunities on their land.

Wind energy leases can also sometimes conflict with existing hunting leases on rural property. Some landowners lease their private land to third parties for hunting as a way to generate some modest additional income. For obvious reasons, the rights of developers under wind energy leases are often at odds with lessees' rights under these hunting leases. *conflicts w/ hunting*

In an effort to address this problem, many wind energy leases include provisions effectively prohibiting hunting activities on the project property, and some require the lessor to represent that there are no hunting leases affecting the land. When a proposed wind farm's land is already subject to a hunting lease, the best option for a developer may be to simply "buy out" the lease, paying a relatively small sum of money to the hunting lessee to prematurely terminate the arrangement.

Fire-related risks

Finally, some neighbors may oppose a proposed wind farm because of the increased fire risk it could bring to their community. From 2002 to 2010, more than 100 wind turbine fire accidents were reported across the world,[113] and fires account for between 9 percent and 20 percent of wind energy insurance claims.[114] According to one manufacturer of turbine fire-suppression systems, the nacelle of a typical utility-scale wind turbine contains up to 200 gallons of hydraulic fluid and lubricants that can quickly ignite when a turbine catches fire.[115] Many other components of the turbine are also highly flammable. If such materials catch flame and fall to the ground under dry conditions, they have the potential to start destructive range and forest fires and threaten nearby homes and other property.

Turbine fires are most commonly ignited by lightning strikes, which is hardly surprising given turbines' tall height and the fact that they tend to be installed at a parcel's higher elevations so that they can capture the most productive winds.[116] The materials used in turbines do not help matters, either—in fact, some have suggested that the carbon fibers increasingly used in commercial wind turbines to make them stronger and lighter also makes them more conducive and more susceptible to lightning strike.[117] Mechanical failures within a turbine's braking systems or other components account for most other fires. A wind project's transformers and other infrastructure can also catch fire on occasion, posing similar risks.

Because wind turbines tower high above the ground and are exposed to the elements, extreme weather other than lightning can also cause turbines to catch fire. In 2001, a highly publicized wind turbine fire occurred in Ardrossan, North Ayrshire, Scotland, during a major storm involving winds of up to 165 miles per hour.[118] The turbine, manufactured by Vestas, apparently had been shut off in anticipation of the storm and was not rotating when the fire broke out.[119] Nonetheless, large pieces of burning debris blew off of the turbine and onto the ground. Fortunately, no significant ground fire erupted as a result.

This incident, which was captured on video and downloaded thousands of times, likely would not have occurred in the absence of such extreme winds. However, multiple wind turbine fires in Germany and the United States occurred in the months following the Scotland fire in far less windy conditions.[120] Although wind turbine fires are relatively rare, developers can do some things to mitigate fire risks associated with wind farms, including proper turbine maintenance and the use of turbine fire suppression systems. Newer turbine models also tend to be less prone to fire.

Conflicts with various other neighboring land uses

Incredibly, the issues described in this chapter constitute only a fraction of the wide array of conflicts that renewable energy developers can encounter with neighboring land uses. For instance, soil erosion resulting from a wind project's access roads or other facilities can conceivably cause stormwater flooding on nearby land. Truck traffic during wind farm construction can permanently damage roads. Wind turbines can even distract travelers on the highway, causing traffic accidents when drivers gawk at the gigantic structures as they pass them along the road. And the following chapter focuses on clashes with one particular category of neighbors—competing wind energy developers—and suggests that disputes with these neighbors require special policy attention in the global effort to generate more clean, renewable energy from wind.

As wind energy development continues its rapid proliferation across the globe, the list of potential conflicts between onshore wind farms and their neighbors will only grow. Landowners, developers, and policymakers should thus be careful to consider unique characteristics or land uses situated near a proposed wind farm when progressing through the planning and development process. Simple thoughtfulness and foresight can help to prevent future problems from arising.

Despite the wide range of conflicts between wind energy projects and neighbors described in this chapter, wind energy's future is not inevitably clouded with one neighbor dispute after another. Local, regional, and national governments throughout the world have crafted a wide variety of laws to address these conflicts. As will be exemplified by the wind energy project described in Chapter 8, successful development in the face of obstacles and challenges is often possible through thoughtful policymaking and diligent developer effort.

Notes

1 *Global Wind Report 2011*, Global Wind Energy Council (2012), available at www.gwec.net/publications/global-wind-report-2/ (last visited Feb. 24, 2014).

2 According to the Global Wind Energy Council, more than 41 GW of wind energy generating capacity were added worldwide in 2011. Less than 40 GW of cumulative capacity had been installed across the globe as of 2003. *See* id.

3 *See* Sarah Azau, *Nurturing Public Acceptance*, WIND DIRECTIONS 30 (September 2011), available at www.ewea.org/fileadmin/emag/winddirections/2011-09/pdf/WD_September_2011.pdf (last visited June 14, 2012) (citing a study available at http://ec.europa.eu/public_opinion/index_en.htm (last visited June 14, 2012).

4 *See The Saint Index: United States, 2011* (2011), available at http://saintindex.info/special-report-energy#windfarms (last visited June 14, 2012).

5 *Wind Energy FAQ*, European Wind Energy Association, available at www.ewea.org/index.php?id=1884 (cited in Azau, *Nurturing Public Acceptance*, WIND DIRECTIONS 34 (September 2011)).

6 Azau, *Nurturing Public Acceptance*, WIND DIRECTIONS 30 (September 2011).

7 Clifford Krauss, *Move Over, Oil, There's Money in Texas Wind*, NEW YORK TIMES A1 (Feb. 23, 2008) (cited in Susan Lorde Martin, *Wind Farms and NIMBYs: Generating Conflict, Reducing Litigation*, 20 FORDHAM ENVTL. L. REV. 427, 445 (2010)).

8 *See, e.g.*, Martin D. Heintzelman & Carrie M. Tuttle, *A Hedonic Analysis of Wind Power Facilities*, 88 LAND ECON 571 (2012) (finding that wind turbine installations studied generally had some measurable adverse impact on the market value of nearby homes). It should be noted that, although there is evidence that some individuals find homes situated near wind farms to be less desirable, much of the research to date on wind energy development's impact on nearby property values has been inconclusive. For a relatively recent summary of this research, *see* Susan Lorde Martin (2010) at 448–49.

9 At least one other writer has described the seeming unfairness of wind energy development from the perspective of neighbors who are excluded from the project. *See, e.g.*, Susan Lorde Martin (2010) at 464–65.

10 *See* GE Energy, *Wind Turbine Facts* at 3 (2009), available at www.ge-energy.com/content/multimedia/_files/downloads/wind_energy_basics.pdf (last visited June 17, 2013) (noting that the "tip height" for a GE 1.5 MW turbine is about 394 feet but that larger onshore turbines can have tip heights of up to 492 feet).

11 Bent Ole Gram Mortensen, *International Experiences of Wind Energy*, 2 ENVTL. & ENERGY L. & POL'Y J. 179, 190–91 (2008).

12 Scottish National Heritage, *Siting and Designing Windfarms in the Landscape* 9 (2009), available at www.snh.gov.uk/publications-data-and-research/publications/search-the-catalogue/publication-detail/?id=1434 (last visited June 18,

2012) (suggesting that the "use of coloured turbines (such as greens, browns or ochres) in an attempt to disguise wind turbines against a landscape backcloth is usually unsuccessful").

13 *Siting and Designing Windfarms in the Landscape 9* (2009).

14 Id.

15 Avi Brisman, *The Aesthetics of Wind Energy Systems*, 13 N.Y.U. ENVTL. L.J. 1, 46 (2005).

16 Id. at 46.

17 Id. at 8 (arguing that aesthetic objections to wind farms are "inconsistent with the aesthetic sensibility we bring to our appreciation of fine art").

18 Id. at 8–9.

19 This is somewhat a generalization, and obviously the common law of nuisance varies by jurisdiction. For more details on the nuances of nuisance law, *see generally* Restatement (Second) of Torts §§§ 821D, 821F, & 822 (1979) (cited in Stephen Harland Butler, *Headwinds to a Clean Energy Future: Nuisance Suits against Wind Energy Projects in the United States*, 97 CAL. L. REV. 1337 (2009)).

20 Butler (2009) at 1350 (citing Mathewson, 395 P.2d 183, 189 (Wash. 1964)); *see also Rankin v. FPL Energy, LLC*, 266 S.W.3d 506 (Tex. App. 2008).

21 See *Burch v. NedPower Mount Storm, LLC*, 647 S.E.2d 879, 893–94 (W. Va. 2007) (cited in Butler (2009) at 1360.

22 Stefanie Huber & Robert Horbaty, *IEA Wind Task 28: Social Acceptance of Wind Energy Projects* (May 2012), available at www.ieawind.org/index_page_ postings/June%207%20posts/task%2028%20final%20report%202012.pdf (last visited June 18, 2012).

23 *IEA Wind Task 28: Social Acceptance of Wind Energy Projects* (May 2012) at 4.

24 Id. at 27.

25 Id. at 26.

26 *Wind Power and the Environment – Benefits and Challenges*, WIND DIRECTIONS at 31 (2006), Available at www.ewea.org/fileadmin/ewea_documents/ documents/publications/WD/wd25-5-focus.pdf (last visited June 12, 2012).

27 *See* Kenneth Kimmell & Dawn Stolfi Stalenhoef, *The Cape Wind Offshore Wind Energy Project: A Case Study of the Difficult Transition to Renewable Energy*, 5 GOLDEN GATE U. ENVTL. L.J. 197, 200 (2011).

28 *See* Dominic Spinelli, *Historic Preservation and Offshore Wind Energy: Lessons Learned from the Cape Wind Saga*, 46 GONZ. L. REV. 741, 748 (2010–11).

29 *See* Katharine Q. Seelye, *Koch Brother Wages 12-Year Fight Over Wind Farm*, NEW YORK TIMES (Oct. 22, 2013).

30 *See* Heidi Willers, *Grounding the Cape Wind Project: How the FAA Played into the Hands of Wind Farm Opponents and What We Can Learn From It*, 77 J. AIR L. & COM. 605, 615 (2012); Erica Schroeder, *Turning Offshore Wind On*, 98 CAL. L. REV. 1631, 1650 (2010).

31 Schroeder, *supra* note 30 at 1650.

32 *See* Timothy H. Powell, *Revisiting Federalism Concerns in the Offshore Wind Energy Industry in Light of Continued Local Opposition to the Cape Wind Project*, 92 B.U. L. REV. 2023, 2034 (2012).

33 *See* Willers, *supra* note 30 at 618.

34 *See* Schroeder, *supra* note 30 at 1651.

35 *See* Dave Levitan, *Will Offshore Wind Finally Take Off on the U.S. East Coast?*, YALE ENVIRONMENT 360 (Sept. 23, 2013), available at http://e360. yale.edu/feature/will_offshore_wind_finally_take_off_on_us_east_coast/2693/ (last visited Oct. 16, 2013).

36 *See generally* E. D. Benson, et al., *Pricing Residential Amenities: The Value of a View*, 16 J. REAL ESTATE FIN. & ECON. 55 (1998).

37 *See* Kara M. Blake, *Marine Spatial Planning for Offshore Wind Energy Projects in the North Sea: Lessons for the United States* at 44, Master's Thesis (University of Washington) (2013), available at https://digital.lib.washington.edu/research works/bitstream/handle/1773/22801/Blake_washington_0250O_11501. pdf?sequence=1 (last visited Oct. 17, 2013).

38 *See* Michelle E. Portman, et al., *Offshore Wind Energy Development in the Exclusive Economic Zone: Legal and Policy Supports and Impediments in Germany and the US*, 37 ENERGY POLICY 3596, 3604 (2009).

39 W. David Colby, Robert Dobie, Geoff Leventhall, David M. Lipscomb, Robert J. McCunney, Michael T. Seilo & Bo Søndergaard, *Wind Turbine Sound and Health Effects: An Expert Panel Review* 3.1.3, American Wind Energy Association and Canadian Wind Energy Association (2009), available at www.windpower ingamerica.gov/filter_detail.asp?itemid=2487 (last visited June 19, 2012).

40 Id. at 3.1.3.

41 *See* Walter Musial & Bonnie Ram, *Large-Scale Offshore Wind Power in the United States* 5, NREL/TP-500-49229 (Sept. 2010), available at www.nrel.gov/ docs/fy10osti/40745.pdf.

42 Colby et al., *supra* note 39 at 3.1.2.

43 Id. at 3.1.2.

44 Id. at 3.1.2.

45 For a useful diagram prepared by General Electric Global Research showing how turbine noise compares to other common noises, visit www.reliableplant. com/Read/27499/How-loud-wind-turbine (last visited Feb. 24, 2014).

46 See Robert Bryce, *The Brewing Tempest over Wind Power*, WALL STREET JOURNAL (March 1, 2010), available at http://online.wsj.com/article/SB100014 24052748704240004575085631551312608.html (last visited June 19, 2012) (noting complaints over turbine noise in Canada, New Zealand, Australia, France, and England).

47 *See, e.g.*, Tom Zeller, *For Those Near, the Miserable Hum of Clean Energy*, NEW YORK TIMES (Oct. 5, 2012), available at www.nytimes.com/2010/10/06/ business/energy-environment/06noise.html (last visited June 19, 2012).

48 Jasper Copping, *Switch-Off for Noisy Wind Farms*, THE TELEGRAPH (Nov. 19, 2011), available at www.telegraph.co.uk/earth/energy/windpower/8901431/ Switch-off-for-noisy-wind-farms.html (last visited June 19, 2012).

49 See Robert Bryce, *The Brewing Tempest over Wind Power* (2010) (mentioning complaints over wind turbine noise in the U.S. states of Texas, Maine, Pennsylvania, Oregon, New York, Minnesota, and Wisconsin).

50 Restatement Second, Torts § 821D (cited in Eric M. Larson, *Cause of Action to Challenge Development of Wind Energy Turbine or Wind Energy Farm*, 50 CAUSES OF ACTION 2d 1, § 16 (2011)).

51 *Rose v. Chaikin*, 453 A.2d 1378 (N.J. Super. Ct. Ch. Div. 1982).

52 Id. at 1380.

53 Id. at 1382.

54 Id. at 1381.

55 See generally *Burch v. NedPower Mount Storm, LLC*, 647 S.E.2d 879 (W.Va. 2007).

56 Id. at 891 (cited in Butler, *Headwinds to a Clean Energy Future* at 1359 (2009)).

57 Nina Pierpont, *Wind Turbine Syndrome: A Natural Experiment* (K-Selected Books, 2009).

58 Id. at 1 (2009).

59 *See* Dr. Pierpont's website, available at www.windturbinesyndrome.com/wind-turbine-syndrome/ (last visited June 19, 2012).

60 A copy of this report is available for download at www.awea.org/learnabout/publications/upload/awea_and_canwea_sound_white_paper.pdf (last visited June 19, 2012).

61 Colby, et al., *Wind Turbine Sound and Health Effects* at ES-1 (2009).

62 For one expert's perspective regarding recent research on this topic, see Jim Cummings, in *Wind Farm Noise and Health: Lay Summary of New Research Released in 2011*, Acoustic Ecology Institute (April 2012), available at www.acousticecology.org/wind/winddocs/AEI_WindFarmsHealthResearch2011.pdf (last visited June 20, 2012).

63 American Wind Energy Association, WIND ENERGY SITING HANDBOOK § 5.4.1 (Feb. 2008), available at www.awea.org/Issues/Content.aspx?ItemNumber=853 (last visited February 24, 2014).

64 WIND ENERGY SITING HANDBOOK § 5.4.1 (2008).

65 Photosensitive epilepsy afflicts only about one in every 4,000 humans. See Graham Harding, et al., *Wind Turbines, Flicker, and Photosensitive Epilepsy: Characterizing the Flashing that May Precipitate Seizures and Optimizing Guidelines to Prevent Them*, 49 EPILEPSIA 1095–98 (2008).

66 Erik Nordman, *Wind Power and Human Health: Flicker, Noise and Air Quality* at 2 (2010), available at www.miseagrant.umich.edu/downloads/research/projects/10-733-Wind-Brief2-Flicker-Noise-Air-Quality2.pdf (last visited June 21, 2012).

67 *See Comment on 'Shadow Flicker' and Photosensitive Epilepsy*, quiet revolution.com (2008), available at www.quietrevolution.com/downloads/pdf/factsheets/C015%20-%20shadow%20flicker%20comment%2003082010.pdf (last visited June 21, 2012).

68 *See Fountainebleau Hotel Corp. v. Forty-Five Twenty-Five, Inc.*, 114 So.2d 357, 359 (1959).

69 At least one other commentator has noted that such a claim would seem to fit outside the scope of common law nuisance. See Butler, *Headwinds to a Clean Energy Future* at 1368–69 (2009).

70 *See generally Burch*, *supra* note 55.

71 For a more detailed description of WindPRO, *see generally* www.emd.dk/WindPRO/WindPRO%20Modules,%20shadow (last visited June 21, 2012).

72 Nordman, *Wind Power and Human Health: Flicker, Noise and Air Quality* at 2 (2010) (describing the prevalence of and range of statutory approaches to addressing shadow flicker by municipal governments in western portions of the U.S. state of Michigan).

73 *See* Press Release: *Wind Turbine Shadow Flicker Study Published* (March 16, 2011), available at www.decc.gov.uk/en/content/cms/news/pn11_025/pn11_025.aspx (last visited June 21, 2012).

74 *See* David Wahl & Philippe Giguere, *Ice Shedding and Ice Throw—Risk and Mitigation*, GE Energy (2006), available at http://site.ge-energy.com/prod_serv/products/tech_docs/en/downloads/ger4262.pdf (last visited June 20, 2012).

75 Jeffrey M. Ellenbogan, et al., *Wind Turbine Health Impact Study: Report of Independent Expert Panel* at 38, Mass. Dept. of Public Protection (January 2012), available, at www.mass.gov/eea/docs/dep/energy/wind/turbine-impact-study.pdf (last visited February 24, 2014).

76 R. Cattin, S. Kunz, A. Heimo, G. Russi, M. Russi, & M. Tiefgraber, *Wind Turbine Ice Throw Studies in the Swiss Alps* (2007), available at www.meteotest.ch/cost727/media/paper_ewec2007_cattin_final.pdf (last visited June 20, 2012).

77 Jeffrey M. Ellenbogan, et al., *Wind Turbine Health Impact Study* at ES 4.3 (January 2012).
78 Wahl & Giguere, *Ice Shedding and Ice Throw* (2006) (citing Temmelin, et al., *Wind Energy Production in Cold Climate* (2007).
79 Kate Galbraith, *Ice-Tossing Turbines: Myth or Hazard?*, NEW YORK TIMES Green Blog (Dec. 9, 2008), available at http://green.blogs.nytimes.com/2008/12/09/ice-tossing-turbines-myth-or-hazard/ (last visited June 20, 2012).
80 Id.
81 Wahl & Giguere, *Ice Shedding and Ice Throw* (2006).
82 Kate Galbraith, *Winter Cold Puts a Chill on Energy Renewables*, NEW YORK TIMES B1 (Dec. 26, 2008), available at http://www.nytimes.com/2008/12/26/business/26winter.html (last visited June 20, 2012).
83 See *NASA's Imagine the Universe!: The Electromagnetic Spectrum,* Introduction, High Energy Astrophysics Science Archive Research Center, Goddard Space Flight Center, http://imagine.gsfc.nasa.gov/docs/science/know_l1/emspectrum.html (last visited Feb. 25, 2011) (comparing and contrasting radio waves, light, microwaves, X-rays, gamma rays, and other types of electromagnetic radiation).
84 WIND ENERGY SITING HANDBOOK § 5.9 (2008).
85 William Kates, *Wind Farms Interfering with Weather Radar in N.Y.*, USA TODAY (Oct. 13, 2009), available at www.usatoday.com/weather/research/2009-10-13-wind-farms-weather-radar_N.htm (last visited June 22, 2012).
86 *See generally* Office of the Director of Defense Research and Engineering, *The Effect of Windmill Farms on Military Readiness* at 52 (2006), available at www.defense.gov/pubs/pdfs/windfarmreport.pdf (last visited Feb. 24, 2014).
87 *See* Donald Zillman, M. E. Walta, & I. D. G. Castiella, *More than Tilting at Windmills*, 49 WASHBURN L.J. 1, 17–18 (2009).
88 Id.
89 Structures that are shorter than 200 feet can sometimes necessitate the filing of a Notice of Proposed Construction when situated near an airport. Roger L. Freeman & Ben Kass, *Siting Wind Energy Facilities on Private Land in Colorado: Common Legal Issues,* 39 COLO. LAW. 43, 49 (May 2010)
90 Id.
91 *See* Kate Galbraith, *Gulf Coast Wind Farms Spring Up, As Do Worries*, NEW YORK TIMES (Feb. 10, 2011) (suggesting that the "Navy would like wind farm construction to stay outside a 30-mile radius of its facilities"). *See also* Donald Zillman, et al., *More than Tilting at Windmills*, 49 WASHBURN L.J. 1, 19 (2009) (stating that, in 2006, the Department of Defense "began issuing notices of 'presumed hazard' to wind project contractors ... for sites within 60 nautical miles of long-range radar installations").
92 Leora Broydo Vestel, *Wind Turbine Projects Run Into Resistance*, NEW YORK TIMES B1 (Aug. 27, 2010).
93 Scott Learn, *Air Force Concerns about Radar Interference Stall Huge Oregon Wind Energy Farm*, OregonLive.com (Apr. 14, 2010), available at www.oregonlive.com/environment/index.ssf/2010/04/air_force_concerns_about_radar.html (last visited June 22, 2012).
94 Id.
95 Juliet Eilperin, *Pentagon Objections Hold Up Oregon Wind Farm*, WASHINGTON POST (Apr. 15, 2012), available at www.washingtonpost.com/wp-dyn/content/article/2010/04/15/AR2010041503120.html (last visited June 22, 2012).
96 Scott Learn, *Pentagon Drops Opposition to Big Oregon Wind Farm*, OregonLive.com (Apr. 30, 2010), available at www.oregonlive.com/environment/index.ssf/2010/04/air_forces_drops_opposition_to.html (last visited June 22, 2012).

 97 *See, e.g.*, Vestel, *Wind Turbine Projects Run Into Resistance* (stating that "many radar systems in use in the United States date back to the 1950s and have outdated processing capabilities—in some cases, less than those of a modern laptop computer"); Elizabeth Burleson, *Wind Power, National Security, and Sound Energy Policy*, 17 PENN. ST. ENVTL. L. REV. 137, 43 (2009) (quoting the United States Department of Energy as stating that "[t]here are a number of technical mitigation options available today, including software upgrades to existing radar, processing filters related to signature identification, [and] replacing aging radar") (citations omitted); Robert Mendick, *Military Radar Deal Paves Way for More Wind Farms Across Britain*, THE TELEGRAPH (Aug. 27, 2011), available at www.telegraph.co.uk/earth/energy/windpower/8726922/Military-radar-deal-paves-way-for-more-wind-farms-across-Britain.html (last visited Dec. 12, 2011).

 98 Robert Mendick, *Military Radar Deal Paves Way for More Wind Farms Across Britain*, THE TELEGRAPH (Aug. 27, 2011), available at www.telegraph.co.uk/earth/energy/windpower/8726922/Military-radar-deal-paves-way-for-more-wind-farms-across-Britain.html (last visited June 22, 2012).

 99 *See, e.g.*, Michael Brenner, *Wind Farms and Radar* 8–9 (2008), Federation of American Scientists, www.fas.org/irp/agency/dod/jason/wind.pdf (last visited March 1, 2011) ("The cost of a single radar installation was said to be in the range of $3–8M, to be compared with the $2–4M cost of a single wind turbine, and the roughly $0.5M annual electric production of a single turbine (5×106 kWh, at $0.10/kWh retail). A wind farm can have hundreds of turbines").

100 John Acher, *Next Thing in Wind Energy: Stealth Turbines*, Reuters (June 29, 2011), available at www.reuters.com/article/2011/06/29/us-windenergy-stealth-vestas-idUSTRE75S2JM20110629 (last visited June 22, 2012).

101 David Danelski, *Mojave Desert: Military Wants to Limit Wind Development*, THE PRESS-ENTERPRISE (May 20, 2012), available at www.pe.com/local-news/topics/topics-environment-headlines/20120520-mojave-desert-military-wants-to-limit-wind-development.ece (last visited June 22, 2012).

102 Mark Colette, *Kingsville Officials Opposed Wind Farm Near Riviera: Navy Says Impact Will Be Minimal*, CORPUS CHRISTI CALLER-TIMES (March 28, 2012), available at www.caller.com/news/2012/mar/28/kingsville-officials-oppose-wind-farm-near-navy/ (last visited June 22, 2012).

103 Rebekah L. Sanders & Ryan Randazzo, *Arizona's Solar Energy Plans Vex Military*, AZCentral.com (Apr. 7, 2012), available at www.azcentral.com/news/articles/2012/03/23/20120323arizona-solar-energy-plans-military.html (last visited June 22, 2012).

104 See *Osage Nation v. Wind Capital Group, LLC*, 2011 WL 6371384 (N.D. Okla. 2011).

105 Id. at 11.

106 Ana Campoy & Stephanie Simon, *Wind Fuels Fight in Oil Patch*, WALL STREET JOURNAL (Nov. 26, 2011).

107 K. K. DuVivier, *Animal, Vegetable, Mineral—Wind? The Severed Wind Power Rights Conundrum*, 49 WASHBURN L.J. 69, 86 (2009).

108 *See* Kathleen D. Kapla & Craig Trummel, *Severing Wind Rights Raises Legal Issues*, NORTH AMERICAN WIND POWER (October 2010), available at http://kaplalaw.com/NAW1010_WindRightsArticle.pdf

109 *See* N.D. Cent. Code 17-04-04 (2010); S.D. Codified Laws 43-13-19 (2010); Wy. Stat. Ann. 34-27-103 (2011); Mont. Code Ann. § 70-14-404 (2011); Neb. Rev. Stat. § 76-3004 (2010).

110 Marty Touchette, *Wind Farm Opponents Cite Concerns for Crop Dusting*, ReviewAtlas.com (Aug. 3, 2010), available at www.reviewatlas.com/news/

x84682246/Wind-farm-opponents-site-concerns-for-crop-dusting?zc_p=0 (last visited June 25, 2012).

111 Id.
112 Paul Monies, *Canadian County Crop Duster Files Lawsuit Against Developers of Planned Wind Farm in Oklahoma*, NewsOK (May 25, 2012), available at http:// newsok.com/canadian-county-crop-duster-files-lawsuit-against-developers -of-planned-wind-farm-in-oklahoma/article/3678347 (last visited June 23, 2012).
113 Scott Starr, *Turbine Fire Protection*, Wind Systems (Aug. 2010), available at http://windsystemsmag.com/article/detail/136/turbine-fire-protection (last visited June 25, 2012).
114 Nancy Smith & Eize de Vries, *Wind and Fire: Reducing the Risk of Fire Damage in Wind Turbines* at 1, Renewable Energy World (Sept.-Oct. 2004), available at www.firetrace.com/assets/pdf/windandfirearticle.pdf (last visited June 25, 2012).
115 Id.
116 Scott Starr, *Turbine Fire Protection*, Wind Systems (Aug. 2010).
117 Nancy Smith & Eize de Vries, *Wind and Fire* at 2 (2004).
118 Donna Bowater & Rowena Mason, *Blown Away: Gales Wreck Wind Turbines and Scottish Storm Wreaks Havoc*, The Telegraph (Dec. 9, 2011).
119 Id.
120 See Sally Bakewell, *Vestas Wind Turbine Catches Fire in Germany, No Injuries*, Bloomberg (Mar. 30, 2012), available at www.bloomberg.com/news/2012-03-30/vestas-says-turbine-catches-fire-at-gross-eilstorf-wind-farm.html (last visited February 24, 2014); Laura DiMugno, *Another Wind Turbine Blaze: Fire Breaks Out at Iowa Wind Farm*, North American Windpower (May 24, 2012), available at http://nawindpower.com/e107_plugins/content/content. php?content.9883 (last visited June 25, 2012); Shaun Kittle, *Wind Turbine Fire Under Investigation*, Press-Republican (Jan. 30, 2012), available at http://pressrepublican.com/0100_news/x897046002/Wind-turbine-fire-under-investigation (last visited June 25, 2012).

3 Wind energy vs. wind energy
Turbine wake interference

Although sailors, kite flyers, and windsurfers have always appreciated windy days, wind has not historically been viewed as a valuable commodity capable of being bought, sold, or stolen. Interestingly, the wind energy industry's dramatic growth over the past decade is beginning to change how society looks at wind resources. Frequent breezes that have done little more than annoy rural landowners for generations are increasingly filling their pockets with cash.

Not surprisingly, as the economic value of wind has grown, the prevalence of disagreements over its ownership has increased as well. For the first time in history, policymakers are facing difficult questions about who owns the wind and are struggling to formulate legal rules that adequately define and allocate wind interests among landowners. Accordingly, this chapter highlights one of the most perplexing types of conflicts that can arise in the context of wind energy development: disputes over rights in the wind itself.

Long wakes, high stakes

The most straightforward way to introduce the concept of competition over wind rights is through a simple example.[1] Suppose that two competing developers have leased adjacent parcels of land for wind energy development. Suppose further that, as shown in Figure 3.1 below, the wind on both parcels tends to blow primarily in a west-to-east direction such that one of the parcels (Parcel U) is effectively upwind of the other (Parcel D). The developer of Parcel U (Upwind Developer) has spent months using anemometers to gather data on wind speeds on Parcel U and has filed a permit application to install a 450-foot-tall, 2.3 MW turbine at Site U. Based on Upwind Developer's data, a project layout that includes a turbine at that site would maximize the productive value of Parcel U's wind resources. The developer of Parcel D (Downwind Developer) has also been measuring wind speeds in preparation to site turbines on that parcel. Based on Downwind Developer's data, a layout with a turbine at Site D would maximize the productive value of Parcel D's wind resources.

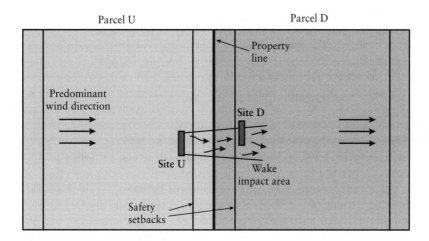

Figure 3.1 Turbine wake interference

Suppose that the proposed turbines at Sites U and D are both eligible for permitting approval because each lies outside a 500-foot "safety" setback applicable in the controlling jurisdiction. These setbacks, which are fairly common minimum setbacks for commercial wind farms, are sufficiently wide to prevent turbine towers from ever falling across property lines.[2] Unfortunately, if the developers each proceeded to install turbines at Sites U and D, the Site U turbine would create a "wake" behind it when rotating that would materially diminish the energy productivity of the turbine on Site D. The two developers involved in this example are thus locked into a simple bilateral conflict over competing rights to capture the energy in the wind currents above their respective parcels.

Wakes like the one at issue in the preceding example can extend for a distance of eight to ten rotor diameters behind a commercial wind turbine— an entire kilometer or more than half a mile.[3] These wakes reduce average wind speeds, which can decrease the energy production of turbines situated downwind.[4] In addition to reducing turbine productivity, such wakes can accelerate turbine "fatigue," causing the gears and machinery in downwind turbines to wear out more quickly than normal.[5]

In western nations, parties in the position of Upwind Developer in the fact pattern above may try to assert that they have superior rights in the wind blowing across their leased property based on the *ad coelum* rule. As described in Chapter 1, the *ad coelum* rule provides that "[to] whomsoever the soil belongs, he owns also to the sky and to the depths."[6] In the wind energy context, one could argue that, based on this rule, ownership of a parcel includes rights to capture the energy in any wind flowing directly above it for productive use.

In contrast, parties in Downwind Developer's position are likely to argue that the wake effects extending onto their property constitute an actionable *nuisance*. Under this theory, Downwind Developer might allege that the Site U turbine's wake unreasonably interferes with its reasonable exercise of its own property rights for wind energy generation. One can recognize the sensibleness of this sort of argument as well, particularly in situations when a neighbor seeks to install a turbine whose wake would substantially interfere with an existing turbine on a downwind parcel.

Turbine wake conflicts like these are often further complicated by the fact that the parties involved in them are frequently at odds with each other even before their dispute arises. Relationships between developers who are competing to develop wind farms in the same community are frequently less than amiable. Even in towns where there are sufficient resources and public support to enable both of two neighboring wind energy projects to succeed, the developers of each project are likely to view each other as arch rivals.

Add to this the unfortunate reality that laws governing wind turbine wake interference are relatively unsettled, and it is easy to recognize the difficult challenge developers can face when such issues arise. Turbine wake interference conflicts are relatively new, so few jurisdictions have clear laws in place to govern them.This dearth of legal guidance can further undermine efforts to peaceably settle turbine wake disputes because there is no statute or solid legal precedent to provide a starting point for negotiations.

Failures to expeditiously resolve turbine wake conflicts can result in costly project delays and litigation to both parties. Rather than risk having to try to protect ambiguous rights in court when faced with such conflicts, some developers may move on to less productive turbine sites, ultimately under-utilizing precious wind resources.

Disputes over wind rights: real world examples

Although many conflicts over wind rights and turbine wake interference go unpublicized, a handful of such conflicts has received media attention in recent years. The following are descriptions of three types of such disputes that have arisen in the United States in the past decade or so and how they were ultimately addressed.

Downwind developers' late calls for wake-based setbacks

On at least two occasions, wind rights disputes have arisen in the state of North Dakota when downwind developers have asked local officials to impose wake-based setbacks on projects situated immediately upwind. In 2005, a downwind developer, enXco, allegedly made such a request in connection with a proposed wind energy project by Florida Power and Light (FPL) in Dickey County, North Dakota. EnXco asked county officials to impose more restrictive turbine setbacks on FPL's upwind parcel to protect the productivity of wind currents flowing onto enXco's own downwind property. The county granted enXco's request and imposed a five-rotor-diameter wake buffer setback requirement on the FPL project. According to at least one source, FPL "abandoned" its wind farm plans relatively soon after the county's decision.[7]

Just three years later, in 2008, a competing downwind developer lobbied the local government to impose wake-based setbacks on another proposed FPL project in North Dakota's Barnes County. Worried that wake effects from FPL's proposed wind farm would reduce the productivity of turbines on its own proposed project, Peak Wind asked the county's zoning commissioners to impose five-rotor-diameter setbacks on FPL's project to prevent such conflicts.[8] Peak Wind argued, among other things that the Public Utilities Commission in the neighboring state of Minnesota had embraced such setback requirements and that FPL had previously complied with them on other wind farm projects.[9] Ultimately, the zoning commissioners refused to impose additional wake-based setbacks on FPL in this instance,[10] but not until both parties had already expended considerable time, funds, and other resources handling the dispute.

General opposition from downwind parties

More general opposition to wind farms based on concerns about wind turbine wakes has also arisen multiple times in recent years. For instance, in 2009, a downwind landowner and developer jointly challenged a draft environmental impact report filed in connection with the Alta-Oak Creek Mojave wind energy project in southern California on this basis. These parties objected to the report in part because it "did not address the impact of wind flow across their property."[11] Although the local planning department dismissed these objections on the ground that wake interference issues were outside the scope of environmental review required under applicable law, the upwind developer had to bear additional costs fending off a wake-based challenge in this case as well.[12]

A different downwind developer also reportedly challenged the expansion of a wind energy project in Umatilla County in the U.S. state of Oregon based on potential wind turbine wake effects.[13] The parties to this dispute eventually resolved it through a confidential agreement, but unquestionably incurred some costs in connection with the dispute.[14]

Patricia Muscarello and wind rights arguments as a NIMBY tool

Even landowners with no evident intention of ever using wind resources on their land may nonetheless use wind rights-based arguments in their efforts to oppose wind farms on neighboring parcels. Landowners using this strategy argue that a wind farm project proposed on lands immediately upwind of theirs would steal their wind resources, even though these landowners are actually more concerned about other impacts that a nearby wind farm might bring.

A landowner in the U.S. state of Illinois has employed this strategy in multiple settings in recent years. Patricia Muscarello is, as one prominent judge described her, "a pertinacious foe of wind farms."[15] She has been particularly active in her opposition to wind projects in close proximity to her own land. On more than one occasion, she has argued in court against nearby wind energy development on the ground that it would deprive her of valuable wind rights.

Muscarello first used her wind rights-based argument in court when challenging the proposed Baileyville Wind Farm, an 80 MW commercial wind energy project on rural land in sparsely-populated Ogle County, Illinois. Muscarello owned nonresidential land adjacent to the area where the developer intended to build the project.[16] In 2005, shortly after the county board issued its first special permit for the wind farm, Muscarello filed a legal complaint against the board and the developer. Among other things, the complaint alleged that issuance of the permit would deprive her of the "full extent of the kinetic energy of the wind and air as it enters her property."[17] Muscarello was seemingly arguing that the board's mere issuance of the permit affected an uncompensated regulatory taking of her wind rights because it would enable the developer to install wind turbines that would reduce the speed of winds flowing onto her land.

The Seventh Circuit Court of Appeals ultimately rejected all of Muscarello's arguments in an issued opinion in 2010,[18] and her later appeal to the United States Supreme Court was likewise denied in late 2011.[19] However, Muscarello's unsuccessful efforts in battling the Baileyville Wind Farm did not dissuade her from making almost identical arguments to challenge new wind energy permitting policies adopted in another Illinois county located less than 30 miles away.

Muscarello owned property in Winnebago County, Illinois, as well and became upset when that county's board amended its zoning ordinance in 2009 to make wind farms a "permitted use" that needed no special use permit. Muscarello opposed these amendments, aimed at easing the permit approval process for wind farms within the county, and filed a legal claim challenging them. Even though no wind farm had yet been proposed in the county since adoption of the amended rules, Muscarello used the same basic wind rights-based claims to challenge these amendments. Among her arguments, all of which failed at the district court level, was the wind-stealing argument referenced above.

On appeal, Judge Richard Posner of the Seventh Circuit noted the incoherent nature of Muscarello's wind rights argument. He acknowledged that "a reduction in wind speed downwind is an especially common effect of a wind turbine" but found it "odd" that Muscarello would object to such wake effects, given her strong displeasure for wind farms. In Posner's words:

> [T]he only possible ... harm that the wind farm could do to her would be to reduce the amount of wind energy otherwise available to her, and the only value of that energy would be to power a wind farm on her property—and she is opposed to wind farming.[20]

Posner aptly identified the inherent flaw in wind rights-based arguments by NIMBY activists. Such landowners are highly unlikely to ever allow productive use of wind on their land, so it would be difficult to justify preventing wind energy development on upwind properties solely to protect these landowners' wind resources. Still, the lack of clear legal rules to allocate wind rights among neighbors can encourage this type of opposition strategy.

Wind rights and turbine wakes: progress to date

Real-life disputes over wind rights and turbine wakes like those just described are almost certain to proliferate as pressure to increase the density of wind energy development escalates in future decades. Enacting laws that clearly and effectively address wind turbine wake interference conflicts can help to prevent these disputes from unnecessarily impeding efficient wind energy development. But what set of rules is best suited to allocate wind rights among neighbors in these contexts? Although wind rights laws are slowly beginning to evolve in some jurisdictions, their development is still very much in the formative stage.

Wind rights and wind severance statutes

Within the U.S., a small handful of states has enacted statutes in recent years aimed at clarifying wind rights ownership to some degree. For instance, a law enacted in the state of Wyoming in 2011 specifically provides that "[w]ind energy rights shall be regarded as an interest in real property and appurtenant to the surface estate."[21] These statutes make clear that rights to capture or make commercial use of wind flowing in the low-altitude airspace above a citizen's land are private property and are not held in public trust or owned by the government or anyone else.

Even simple, clarifying statutes like the one enacted in Wyoming can help to reduce the legal uncertainty associated with wind energy leasing and are thus laudable on that account. In some nations, the concept that private landowners generally hold interests in wind resources above the surface of their land is less clear. One such nation is China, which has been among the world leaders in wind energy development for years.[22] At least one regional government entity in the country has interpreted Article 9 of China's constitution to provide that wind and solar energy are state-owned resources and thus are not private property.[23] Although no other major government entity in China has publicly taken this stance, the mere existence of such a controversy can potentially damper wind energy investment. Basic wind rights laws like those in Wyoming fully eliminate such questions.

New laws passed in many U.S. states also make clear that wind rights cannot be "severed" and transferred separately from surface rights.[24] Landowners in the United States have legitimately severed mineral rights from surface rights for decades, so when the concept of wind rights emerged some citizens began attempting to sever these rights as well. In fact, at least one court in the United States has held that wind rights severance is permissible in some contexts.[25]

Severing wind rights could seemingly be beneficial in some situations. For example, it would allow a landowner who has leased rural ranchland for wind energy development to convey the land to one child and the wind rights (and accompanying rental income stream) to a different child. However, because of complications that could result from allowing such severance, all U.S. states that have enacted statutes relating to wind rights severance have taken the approach of prohibiting landowners from severing their wind rights from the surface estate. On the whole, these laws are beneficial in that they provide at least some additional certainty on the issue of wind rights severance, potentially reducing legal risks for wind energy developers.[26]

Unfortunately, none of the new statutes just described resolves the conflict described earlier in this chapter when a wind turbine's wake crosses a property line and adversely impacts the turbine of a downwind neighbor. This question largely remains unanswered and continues to generate costly uncertainty for wind energy developers in most jurisdictions around the world. What sorts of rules are best equipped to address these perplexing and increasingly common disputes?

Analogies to other natural resource laws

Laws governing other natural resources are a logical place to begin when considering possible rules to govern wind turbine wake interference disputes. It is thus hardly surprising that multiple commentators have explored possible analogies to these other sets of legal rules when analyzing wake interference issues. Sadly, none of these related bodies of law seems to provide a rule regime adequately tailored to resolve conflicts over wind rights. The following excerpt from one of the author's articles on this subject compares and contrasts wind with water, minerals, and other natural resources and discusses why laws governing these resources are not well suited for wind.

> Existing natural resource law is a logical starting point for crafting rules
> to govern wind rights. Longstanding laws already allocate interests
> in water, wild animals, minerals, airspace, and subsurface oil and gas

among landowners. Legal scholars have drawn comparisons between wind and these other resources for decades.

In a sense, wind is like a wild animal: to the extent that the energy in wind is "captured" by a landowner's turbine, that energy never crosses the property line for potential capture by landowners immediately downwind. For centuries, common law rights in many wild animals have been allocated based upon the "rule of capture," a doctrine famously highlighted in *Pierson v. Post* that gives title in wild animals to the first party who kills or captures them. Under old English law, landowners generally had constructive possession of wild animals that were located on their property. Landowners in the U.S. have no possessory rights in wild animals upon their land but can typically preclude others from hunting there, giving them an exclusive right to capture animals and claim title. A rule of capture for wind would arguably give landowners the exclusive right to capture wind crossing their property, without liability to downwind neighbors for consequent reductions in wind flow.

Wind also bears similarities to subsurface oil and gas. Both oil extraction and wind energy development require significant capital investment, and both can potentially reduce the amount of valuable resources available to adjacent property owners. Oil and gas law is rooted in the rule of capture and in another foundational principle of natural resources law known as the *ad coelum* doctrine. The *ad coelum* doctrine provides that "to whomsoever the soil belongs, he owns also to the sky and to the depths," meaning that a fee owner of a parcel of land owns all of the minerals, oil, and gas immediately beneath the parcel's surface and all the usable airspace immediately above it. Nearly a century ago, state courts in Texas and other states essentially combined the rule of capture and *ad coelum* doctrine to develop a rule of capture to oil and gas that entitled landowners to as much subsurface oil as they could extract from directly below their properties.

Not surprisingly, applying a pure rule of capture approach to oil and gas rights ultimately proved to be problematic. Subsurface oil is typically in liquid form and exists in large pools that span beneath numerous parcels, so extraction occurring on one parcel can often reduce the volume of oil remaining under other parcels. The rule of capture thus led to wasteful drilling practices because landowners operating under the rule had an incentive to extract oil as quickly as possible to avoid forfeiting it to neighbors. Over time, state legislatures have addressed these problems by enacting more sophisticated unitization statutes and other laws structured to encourage more efficient oil extraction. While unitization statutes are effective at allocating oil and gas rights among large groups of landowners, such approaches would be far less apt at handling the typical two-party wind rights dispute.

Wind is also analogous to water flowing in rivers and streams across the surface of land. Unlike oil, water often *crosses* property rather than residing in stationary pools beneath it. Landowner conflicts over water rights thus exhibit a directional (upstream/downstream) element comparable to the upwind/downwind nature of wind.

Laws in many [U.S.] states allocate interests in surface water among landowners based upon the doctrine of riparian ownership, which provides that landowners whose properties abut a water source possess rights to make "reasonable use" of water from the source in proportion to their respective parcels' frontage on the lake, river, or stream. Of course, since all of a parcel's surface area "fronts" the airspace through which breezes blow, analogies to the riparian ownership doctrine are unlikely to yield a workable rule for wind rights.

In lieu of riparian ownership laws, some states with more arid climates have embraced the prior appropriation doctrine. Under the prior appropriation doctrine, the first party to make beneficial use of water from a natural water source thereby acquires rights to continue using that amount for that purpose. One could certainly imagine a "first-in-time" rule for wind rights resembling the prior appropriation doctrine under which the first landowner to install a wind turbine (and thereby make beneficial use of the wind) could claim rights in the affected wind flow against the competing interests of neighbors. However, a first-in-time approach would arguably contradict the *ad coelum* doctrine by effectively reallocating limited rights in airspace to initial turbine installers. Such an approach would also lead to productive inefficiency whenever the first of two competing turbine sites to be developed had less valuable wind resources than the undeveloped site.

(Troy Rule, *Sharing the Wind*, THE ENVIRONMENTAL FORUM, Vol. 27, No. 5, pp. 30–33 (September/October 2010))

A first-in-time rule?

Interestingly, at least one local government has effectively adopted a "first-in-time" rule for wind rights that is comparable to the prior appropriation rules for water rights described in the excerpt above. A provision adopted in Otsego County, New York, requires that "a 'wind access buffer' equal to a minimum of five (5) rotor diameters ... be observed from any existing off-site wind turbine generator tower" for all newly-proposed wind turbines.[27] This rule effectively allocates competing wind rights to whoever is the first to make beneficial use of wind resources by erecting a wind turbine on their land.

Consider, for instance, the effect of such a "first-in-time" rule in the hypothetical fact pattern set forth in connection with Figure 3.1 above. In that scenario, if Downwind Developer had already installed a turbine at Site D when Upwind Developer filed its permit application for a turbine at Site

U, Upwind Developer's application would have violated the wind access buffer and been denied on that ground. In contrast, if Upwind Developer had been the first to install a turbine, Downwind Developer would have been prohibited under the rule from installing a turbine at Site D because of its close proximity to the existing turbine at Site U.

Regrettably, one disadvantage of using a first-in-time rule for wind turbine wake interference is its potential to promote inefficient "strategic behavior" problems among developers. For example, suppose that a first-in-time rule applied in Downwind Developer's jurisdiction but that the developer was not prepared to construct a significant wind farm on Parcel D anytime soon. In such a situation, Downwind Developer might elect to install multiple *small*, inexpensive wind turbines along its common boundary line with Parcel U as a way of securing wind rights in the area for future development. Once those small turbines were installed, Upwind Developer would be unable to site any turbines—large or small—within five rotor diameters of them. For obvious reasons, such opportunistic behavior could create inefficiencies and perverse incentives in the context of wind energy development.

The uniqueness of wind

In summary, laws governing oil, water or wild animals are not well-suited for allocating rights in wind because they do not account for wind's distinct characteristics. To continue quoting from the article referenced above:

> Wind is truly unlike any other natural resource. It is ephemeral and invisible. Unlike oil, minerals, animals or water, wind is not easily transported or diverted for use elsewhere.
>
> Wind also differs from other natural resources in that its productive value is often location-specific. If one million barrels of oil reside in a large subsurface pool, then ultimately about one million barrels will be extracted and added to the world's oil supply regardless of whether the extraction occurs on a particular parcel or on neighboring parcels. Similarly, the total volume of usable water in a river is roughly constant regardless of whether water is diverted and used by an upstream landowner or by someone located miles downstream. But such is not the case for wind. Higher average wind speeds generate far more electric power, and average wind speeds are influenced by topography and can vary significantly by location. Thus, the amount of wind energy generated from a given set of properties is often based on *where* upon those properties wind turbines are installed. Maximizing wind energy production from a fixed set of properties and turbines requires installing the turbines in those locations that collectively exhibit the greatest wind energy potential. The impact of *where* wind is captured on *how much* wind power is generated makes wind unique among natural resources.

Not only is wind highly unique among natural resources, but disputes over wind rights are also materially different from any other conflict covered in this book. Wind turbine wake interference conflicts do not involve tradeoffs between renewable energy development and some other land use such as a residential neighborhood, wildlife habitat, or scenic territorial views. Instead, they pit one renewable energy developer against another in clashes over the same type of development, often undermining efficiency on both sides of the property line.

If analogies to water law, oil and gas law, or laws governing rights in wild animals do not provide satisfactory legal rules for wind rights, what rules *are* best suited for allocating wind rights among landowners? Unfortunately, as described in the following subsections, the basic policy strategies for dealing with turbine wake interference that have been adopted to date frequently fall short of the basic goal of maximizing the productivity of the wind resources involved.

Wake-based setbacks?

Other than Otsego County's "first-in-time" approach highlighted above, the only significant policy response to the turbine wake interference problem to date has been to impose wide, wake-based setbacks to prevent such conflicts from arising. The U.S. state of Minnesota has embraced this approach. In that state, commercial wind energy developers must site turbines at least five rotor diameters away from all boundaries of their project control area on the predominant wind axis and three rotor diameters from boundaries on the secondary wind axis unless they receive special government approval to do otherwise.[28] Given that rotor diameters for many modern commercial wind turbines are upwards of 100 meters, these setbacks essentially prohibit turbine installations for a half kilometer behind a turbine. Although these setbacks certainly go a long way in preventing wake interference conflicts, they can also be sources of significant inefficiency, taking vast sections of prime wind energy land out of productive use.

To illustrate the disadvantages of using setbacks to govern wake interference problems, let us return to Parcels U and D described earlier in the chapter. For simplicity, assume that these are the only two parcels in the relevant vicinity that have any developable wind resources. Suppose further that, taking wind speed data and potential wind turbine wake interference issues into account, the overall turbine layout that would maximize the aggregate energy productivity of the wind resources on these two parcels is as shown in Figure 3.2. If a single developer held development rights covering both parcels, the developer would choose this layout to maximize the productivity of the wind resources involved.

An optimal set of legal rules for allocating wind rights would result in this Figure 3.2 layout, regardless of whether the parcels had been leased together by a single developer or separately by two competing developers.

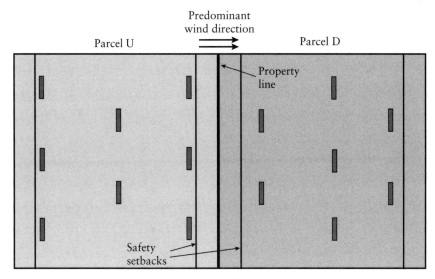

Figure 3.2 Optimal turbine layout for parcels U and D, designed by a single developer who held rights in both parcels

Under this layout plan, the total site would host 15 turbines in all, spaced so as to minimize wake interference while maximizing energy productivity. Eight of the turbines would be on Parcel U, which would allow for only seven turbines on Parcel D. Of course, the fact that Parcel D would have one fewer turbine than Parcel U would be of no concern to a single developer who was developing both parcels together.

Suppose instead, however, that Parcels U and D were separately owned and that two competing wind energy developers, Upwind Developer and Downwind Developer, were independently planning to develop wind farms on their respective parcels. Under this scenario, each developer would likely focus primarily on how to maximize the energy productivity of wind resources on its own parcels, with little or no regard for impacts on parcels outside its project area. Under that set of incentives, the developers' project layouts might resemble those in Figure 3.3. Each developer would squeeze a total of eight turbine sites onto its respective parcel.

This layout is different from the one shown in Figure 3.2 in that Downwind Developer in this diagram has arranged turbine sites on Parcel D to fit an eighth turbine onto that parcel. This layout would allow Downwind Developer to fit an additional turbine on its parcel, but it is also likely to create a wind turbine wake interference conflicts involving Downwind Developer's westernmost three turbines.

In this scenario, Upwind Developer would likely assert that its layout satisfied existing permitting requirements and thus should not create any liability for downwind wake effects. In contrast, Downwind Developer

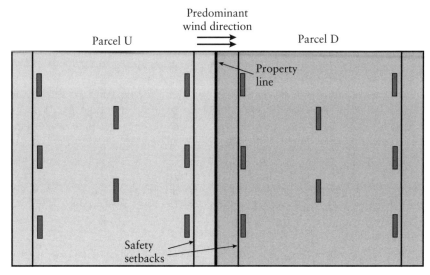

Figure 3.3 Hypothetical turbine layouts for parcels U and D when designed by
two independent developers

would claim that Upwind Developer's layout unreasonably interfered with
Parcel D's wind rights and might seek an injunction to prevent Upwind
Developer from using its layout or damages for the losses resulting from it.

In Minnesota, this sort of conflict would not arise at all. In that state,
"wind access buffer" setbacks of five rotor diameters on the predominant
wind axis and three rotor diameters on the secondary axis would disallow
both developers' proposed turbine layouts.[29] Such setbacks are relatively
effective at preventing turbine installations in locations that are close
enough to parcel boundaries to generate wake-related disputes.

However, these sorts of well-intended setback restrictions can also greatly
reduce the amount of developable wind energy land within a jurisdiction.
The potential inefficiency of this approach is illustrated in Figure 3.4. A
wake-based setback imposed on the relatively small parcels used in the
fact pattern above would require both of the developers in the fact pattern
to scale back their turbine layouts to just five turbines each—a 50 percent
aggregate reduction in aggregate wind energy generating capacity from what
would have been possible under the Figure 3.2 configuration. Even when
only one or two optimal turbine sites are precluded due to such setback
rules, the resulting inefficiencies are often likely to exceed those that would
have resulted under the absence of a setback rule. At least one renewable
energy developer has commented that such a setback requirement would
threaten the economic feasibility of its wind project.[30]

Given the significant inefficiencies associated with wake-based setbacks,
it is hardly surprising that multiple groups expressed disapproval when
the Minnesota Public Utilities Commission first considered imposing its

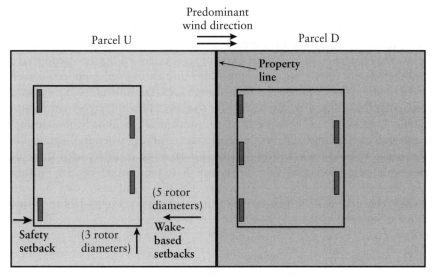

Figure 3.4 Hypothetical turbine layout for parcels U and D in a jurisdiction with wake-based turbine setback requirements

aggressive wake setback rules. Among these opposition groups were 17 participants and supporters of community-based energy development within the state who argued that the setbacks were "overly conservative" and did "not economically or efficiently utilize state wind resources."[31] Regrettably, the Commissioners ignored these observations and adopted the setback rules, unnecessarily precluding wind energy development on thousands of acres of developable land in that state.

Other plausible strategies

Besides wake-based setbacks, what other potential policies could govern disputes over turbine wakes? At least three other basic policy strategies for addressing wind turbine wake interference conflicts are worth mentioning, some of which seem more promising than others.

Compulsory unitization?

In recent years, some policymakers in the United States have advocated rules for wind energy development that are modeled after compulsory unitization laws applicable in some states to subsurface oil and gas. These rules would treat a potential wind farm development area and those parcels in its vicinity as a large "pool" or cooperative. A wind energy developer would be obligated to compensate all landowners within such areas, regardless of whether turbines were ultimately installed on their land. Unlike laws requiring payments to communities that host wind farm neighbors to help prevent local resistance based on aesthetic or other general NIMBY concerns, this policy approach would be aimed at compensating landowners for impacts on their wind resources.[32]

In 2011, legislators in North Dakota in the United States introduced a bill that would have required compulsory unitization for commercial-scale wind energy development in that state.[33] All landowners whose "wind resources are affected by the siting of a commercial wind energy conversion facility" would have been entitled to compensation under the statute.[34] Although the bill was never enacted, the idea of using unitization-like structures to avoid wind turbine wake interference conflicts with those on the fringe of a wind energy project is one that could resurface in the future. Unfortunately, compulsory unitization can add significant costs to wind energy development, which is why some industry stakeholders in North Dakota resisted it.[35]

Wake interference as a nuisance?

Another possible way of addressing turbine wake interference conflicts would be to categorize wakes as an actionable nuisance whenever they materially adversely affect a downwind neighbor's wind farm. A turbine wake that reduces energy generation at a downwind wind farm is arguably a substantial non-trespassory interference with a productive use of neighboring land and thus fits the basic requirements for a nuisance claim. In that sense, a court could conceivably determine that turbine wake effects are not altogether different from foul odors, loud noises, or other common law nuisances. In the United States, courts seeking to apply nuisance law to such conflicts are likely to try to "balance … the costs and benefits of the land use claimed to have caused a nuisance"—a wind turbine in this case. Under this sort of analysis, a turbine's wake might theoretically give rise to an actionable nuisance in those instances when the magnitude of the wake's harm to the downwind landowner exceeded any benefits of siting the offending turbine in its particular location on the upwind property.[36]

Laws recognizing turbine wake interference as a nuisance would arguably be better than the current lack of *any* legal rule in that such laws would help to reduce legal uncertainty associated with these conflicts. Once legislators enacted such a rule, landowners would know that they could potentially be liable for wake effects and could design their wind energy project layouts accordingly.

On the other hand, a rule recognizing nuisance claims for turbine wakes would implicate its own problems and would arguably diverge from existing law. For starters, commercial wind turbines stand hundreds of feet in the air, and wind farms are still a fairly unusual type of land development in most areas. One could thus argue that commercial wind energy development is a hypersensitive land use—not an ordinary land use deserving of nuisance law protection. Also, laws recognizing wake effects as a nuisance would unnecessarily invite litigation over these conflicts. Downwind landowners would have standing to sue and employ scarce judicial resources whenever an upwind neighbor installed a wind turbine that reduced wind productivity on their land.

A clear non-nuisance rule?

A third potential way of governing wind turbine wake interference conflicts would be to adopt the exact opposite of a nuisance-based approach: a simple rule providing that wind turbine wake interference can *never* give rise to any legal claim. Under this straightforward rule, landowners would be free to capture any wind flowing above their land without liability to neighbors for downwind wake effects.

Like the nuisance rule, this "non-nuisance rule" would greatly clarify how wind rights are legally allocated among neighbors, reducing legal uncertainty that might otherwise deter efficient wind turbine siting. Such a rule would also arguably be more in line with the *ad coelum* doctrine because it would allocate wind rights strictly based upon surface ownership, solely protecting rights to whatever wind resources happened to flow in the airspace directly above any given parcel. Because lawsuits based on wind turbine wake interference would be disallowed under this sort of rule, the rule would even generate far less litigation than would a nuisance-based approach.

Particularly in jurisdictions where the terrain is very flat and there is minimal local variability in wind resources, this non-nuisance rule may be the best policy approach for dealing with turbine wake conflicts for the reasons just mentioned. There is reason to believe that this rule would have some capacity to promote optimal turbine layouts, even when two neighboring, competing developers are involved.

The potential merits of the non-nuisance rule are easier to understand through a revisiting of the fact pattern set forth earlier in the chapter involving Figures 3.2 and 3.3. Suppose, in that scenario, that Upwind Developer was the first to submit its layout for Parcel U—an eight-turbine layout like those shown on Parcel U in Figures 3.2 and 3.3. If the relevant jurisdiction had adopted a non-nuisance rule for turbine wakes, Downwind Developer would have known that it had no legal protection against any wake effects affecting Parcel D when this Parcel U layout plan came to light. Accordingly, Downwind Developer would have revised its Parcel D layout from the eight-turbine version in Figure 3.3 to the seven-turbine design shown in Figure 3.2—the optimal outcome.

The non-nuisance rule would be slightly less effective whenever the downwind party involved is the first to develop its project. Returning to the same fact pattern, suppose instead that Downwind Developer had begun siting a wind energy project on Parcel D before Upwind Developer began siting work on Parcel U. In a jurisdiction with a non-nuisance rule,

Downwind Developer in this situation would recognize the possibility that wind turbines later installed on Parcel U could ultimately create wakes that interfered with turbines on the westernmost portions of Parcel D. Unwilling to merely assume this risk, Downwind Developer would be faced with a decision. It could either refrain from developing those westernmost areas or it could contact Upwind Developer and try to negotiate a deal to resolve the issue. Specifically, Downwind Developer could offer to purchase an easement or covenant prohibiting the installation of any wind turbines on Parcel U capable of creating wakes that could harm turbines in Downwind Developer's proposed wind farm array.

Although the non-nuisance rule would provide legal certainty capable of helping parties to reach voluntary agreements in such situations, other common obstacles could easily prevent some parties from ever reaching such arrangements. Developers in these contexts may often be arch rivals and thus be hesitant or even unwilling to negotiate. In cases where a voluntary arrangement fails to occur for this or any other reason, parties in the shoes of Downwind Developer may ultimately conclude that installing turbines in areas that are vulnerable to wake interference areas is too risky. In our fact pattern, this could lead Downwind Developer to elect not to install turbines in such areas and instead site them only in the five easternmost turbine sites of Parcel D, as shown on Figure 3.5. Such a decision would lead to a total of only 13 turbine installations on Parcels U and D—two less than under the optimal layout. In summary, whenever the parties in these situations are

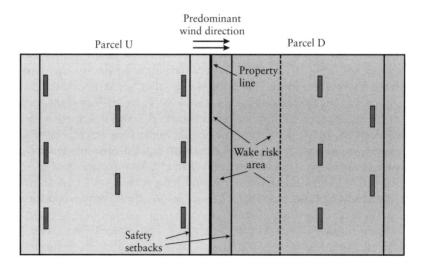

Figure 3.5 Potential turbine layout for parcels U and D under a non-nuisance rule if downwind developer develops first

unable to strike a deal, a non-nuisance rule still has the potential to lead to inefficient outcomes.

For similar reasons, a simple non-nuisance rule can also lead to sub-optimal wind turbine siting in regions where wind resources vary significantly by location. In many areas of the world with hilly terrain or wind "corridors," such as the Columbia Gorge corridor in the northwest United States or the southern Wyoming corridor in intermountain western portions of the United States, wind resource quality can vary greatly over short distances.[37] In these regions, micro-siting turbines in the precise locations where wind resources are particularly strong is crucial to maximizing overall energy productivity in wind farm design. The following numerical examples help to illustrate how a non-nuisance rule can potentially interfere with the efficiency of this micro-siting in these regions with location-sensitive wind resources.

Suppose, for example (as "Scenario #1"), that Parcels U and D are situated in an area of uneven terrain and a high degree of locational wind speed variability. For simplicity, assume that each developer plans to install only one turbine on each parcel. Based on wind speed data, the wind resources at Site U_1 are worth $300,000. Those at Site D_1, which is at a lower elevation, are worth $200,000. However, as illustrated in Figure 3.6, if the developers install both turbines, a wake from the Site U_1 turbine would diminish the energy productivity of and ultimately damage the turbine at Site D_1, reducing the site's value to only $125,000.

Each developer has also identified the best alternative turbine site on its respective parcel. These sites, labeled Sites U_2 and D_2 on Figure 3.6, would

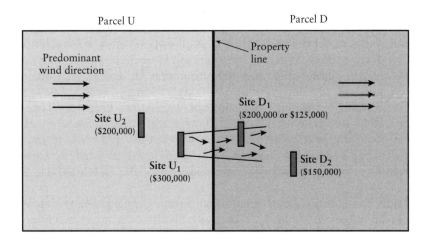

Figure 3.6 Potential turbine sites on parcels U and D: scenario #1

generate significantly less power than their first-choice sites, having values of only $200,000 and $150,000, respectively.

An optimal wind rights law would result in the two turbine installations on these parcels that would maximize the productive value of the wind resources at issue. In this case, those two sites are Site U_1 and Site D_2. Turbines on these sites would produce energy with an aggregate value of $450,000—a combined total that exceeds any other possible pair of turbine installations.

A non-nuisance rule is likely to produce the result under the assumptions of this example. Specifically, the rule would entitle Upwind Developer to install its turbine at Site U_1 without liability and Downwind Developer would be unwilling to pay enough money to convince Upwind Developer to do otherwise. Suppose that Downwind Developer, recognizing the potential for wake interference at Site D_1 if a turbine is installed at U_1, were to approach Upwind Developer and attempt to strike a deal to prevent a turbine installation at U_1. $50,000 (the difference between the values of Sites D_1 and D_2) is the largest sum of money that Downwind Developer would be willing to pay to compensate Upwind Developer to install its turbine at Site U_2 rather than Site U_1 and prevent wake interference. Upwind Developer would be unwilling to accept that amount because siting at U_2 instead would cost Upwind Developer $100,000 (the difference between the values of Site U_1 and U_2). The parties would thus be unable to reach an agreement and Downwind Developer would install its turbine at Site D_2 to avoid wake interference from Upwind Developer's turbine at Site U_2—the socially optimal outcome.

Unfortunately, the non-nuisance rule performs less well where the more valuable wind resources are on the *downwind* parcel rather than on the *upwind* parcel. Suppose, for example (as "Scenario #2") that all assumptions remained the same as in Scenario #1 above except that Site D_1 was actually at a higher elevation with greater wind speeds and was thus twice as valuable as under the prior scenario. As shown in Figure 3.7, maximizing the productive value of the wind resources at issue under this revised set of facts would require turbine installations at sites U_2 and D_1—an outcome resulting in sites worth a combined $600,000. A non-nuisance rule would be less certain to produce that optimal result than under Scenario #1 because the rule would legally entitle Upwind Developer to install a turbine on its own most productive site, U_1. Downwind developer would thus be at the mercy of Upwind Developer to negotiate a deal whereby Upwind Developer relocates to Site U_2, and attempts to strike such a bargain may fail.

Again, this idea is easier to see through a more detailed explanation using the dollar values in Figure 3.7. The only plausible means of reaching the optimal outcome under Scenario #2 would be a voluntary agreement in which Downwind Developer paid Upwind Developer between $100,000 and $150,000 to install its turbine at U_2 rather than U_1. Unfortunately, even

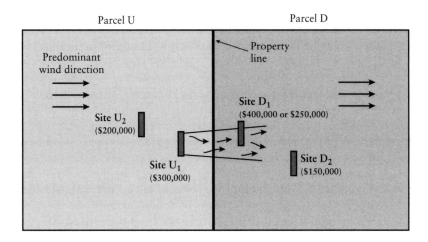

Figure 3.7 Potential turbine sites on parcels U and D: scenario #2

though this agreement would be in the best of interest of both parties and of society, it is uncertain at best that it would actually occur. As described earlier in the chapter, competing wind energy developers often view each other as rivals. Voluntarily negotiating to resolve turbine wake interference issues thus may be quite difficult for them, even if the laws governing such interference are made clear under a non-nuisance rule. If bargaining does fail, inefficient turbine siting decisions (in this case, installations at Sites U_1 and D_2) can result.

In summary, neither the existing approaches for addressing wind turbine wake interference conflicts nor any of the most obvious potential legal rules to govern them are very satisfactory, because all are likely to lead to sub-optimal wind farm siting decisions in some cases. Given the strong public policy interest in not wasting scarce wind resources, a more sophisticated wind rights law may be required that is better equipped to promote productive efficiency in these unique situations.

A more suitable alternative: the "option approach"

Technological innovation sometimes necessitates policy innovation, and such may be the case as policymakers seek for better ways to address the sorts of wind rights disputes already highlighted in this chapter. The following is a description of a unique set of rules for addressing conflicts over wind rights that is arguably better tailored to wind's unique characteristics. This two-part policy approach is designed to respect existing property interests in the scarce resources involved in wind energy development while still giving downwind owners a sure-fire means of acquiring protection against wake interference whenever the wind resources on their land are superior to those on upwind parcels.

The first part of this approach would be a simple non-nuisance rule, providing that landowners are never liable for downwind wake effects caused by turbines installed on their properties. Except in cases where an upwind defendant was clearly creating wake effects out of spite, landowners would be free to install turbines on their parcels without having to worry about being sued for the impacts of their turbines' wakes on properties downwind.

To provide a means for downwind landowners to obtain wake protection in cases where their lands have comparatively stronger wind resources, the second part of this policy approach would create the equivalent of a purchase option for an easement in favor of downwind parties. This policy would require landowners to send written notice to downwind neighbors before installing any turbines situated within five rotor diameters of a property line along the predominant wind axis or three rotor diameters along the secondary wind axis. These neighbors would then have an opportunity to purchase from the upwind landowner the equivalent of an easement or covenant preventing the siting of turbines within those setback distances. Government officials would seek to set the purchase price for this wake interference protection at an amount equal to the expected loss that the upwind party would incur from the resulting restriction. Faced with such a price, a rational downwind neighbor would elect to pay the amount and secure wake interference protection only if its expected loss from wake interference exceeded the upwind party's expected losses under the restriction.

A numeric example

Given the relative complexity of this rule, it is admittedly difficult to grasp it without a concrete example. To help better illustrate this option-based policy and how it might function in practice, let us return once more to the hypothetical fact pattern and diagrams involving Upwind Developer and Downwind Developer used earlier in the chapter. Assume now, however, that the county government where the parcels are located has adopted an ordinance establishing wake-based turbine setbacks in areas zoned for commercial wind energy development. These five-rotor-diameter by three-rotor-diameter setbacks are sufficient to prevent any substantial wake effects on adjacent properties downwind. However, these particular wake setbacks are different from those imposed in Minnesota and some other places in that they are *waivable* upon the satisfaction of certain conditions prescribed in the ordinance.

Suppose that Upwind Developer has filed a permit to install a new wind turbine at Site U_1, a site well within this waivable wake setback area, as shown on Figure 3.8. Under the county's ordinance, Upwind Developer would be required to send written notice to Downwind Developer in connection with its permit application for a turbine at this site. Downwind Developer would then have an option to pay Upwind Developer to make the setback permanent and thereby prevent installation of the Site U_1 turbine. County officials would seek to set the price of this option at an amount equal to Upwind Developer's expected loss from being prohibited from installing turbines within the wake setback area. Information from the Upwind Developer's own wind studies and evidence about prevailing wind energy prices would aid county officials in setting this price.

Suppose that the wind resources on Parcels U and D have the values reflected in parentheses in Figure 3.8 and that, for simplicity, each developer can install only one turbine on its respective parcel. Note in Figure 3.8 that the value of Site D_1 is either $400,000 or $250,000, depending on whether it is subjected to wake interference.

Based on these figures, a rational party in Downwind Developer's position would be willing to pay up to $150,000 to make the waivable setback on Parcel U permanent and thereby be protected from wake interference. So long as the county's estimate of the loss to Upwind Developer under the setback did not exceed $150,000, Downwind Developer would pay the stipulated amount and the setback would become permanent, preventing Upwind Developer from installing a turbine at Site U_1. A "Notice of Permanent Wake Setback" would be recorded against the upwind property, and turbines would ultimately be installed on the more productive,

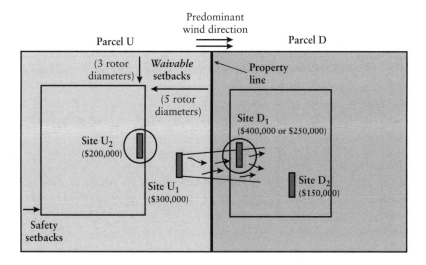

Figure 3.8 Effect of waivable setbacks under an option-based policy (scenario #1: best wind resources situated on *downwind* site)

downwind site (D₁) and at Site U₂. This is the optimal siting scheme under the given set of assumptions, having a combined value of $600,000—more than any other possible pairing of upwind and downwind sites.

This option-based policy would also produce the optimal siting arrangement if the best wind resources at stake were on Parcel U, the *upwind* parcel, instead of on Parcel D. To illustrate this important point, suppose that Site D₁ were instead worth only $200,000 without wake interference and $125,000 with interference, as shown on Figure 3.9 below. If all other assumptions remained the same, Downwind Developer would be willing to pay no more than $50,000 to make the waivable setback permanent under this scenario because it would only lose that amount if it sited its turbine at Site D₂ instead of at Site D₁. So long as the local government's determined valuation of the upwind site was at least $50,000, Downwind Developer would thus elect not to exercise its option to make the waivable setback permanent. The local government would record a "Notice of *Waiver* of Wake Setback" to provide constructive notice that the relevant wake setback had been waived, and turbines would ultimately be installed at Sites U₁ and D₂—the most productive siting scheme under this revised set of facts.

In both of the above-described scenarios, the use of waivable wake setbacks would enable the parties to peaceably resolve their turbine wake interference conflicts while maximizing the productive value of the wind resources involved. Landowners operating under such a policy would still be free to reach voluntary agreements to settle siting disputes. However,

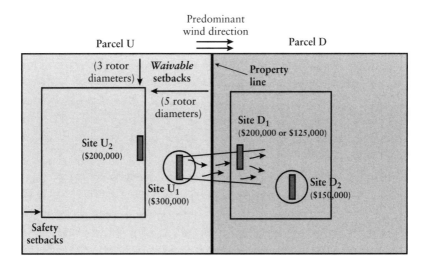

Figure 3.9 Effect of waivable setbacks under an option-based policy (scenario #2: best wind resources situated on *upwind* site)

whenever voluntary bargaining failed for any reason, downwind owners would have an alternative means of acquiring protection against wake interference if they were willing to purchase it for an amount equating to its fair value. This policy would not only reduce legal uncertainty regarding the allocation of wind rights among landowners; it would also promote the overarching policy goal of promoting productive efficiency in wind turbine siting.[38]

As of yet, no local government has begun using waivable wake-based setbacks to address the issue of wind turbine wake interference conflicts in its jurisdiction. As the value of wind resources continues to grow and such conflicts become increasingly common, hopefully governments in places where wind resources are highly location-specific will not be afraid to consider such a policy approach. Although it is a bit more complex than a pure non-nuisance rule, this sort of two-part rule has the potential to better balance the competing goals of fairness and efficiency in the wake interference context.

Notes

1 Portions of this chapter are adapted from the author's other published writings on this topic, including Troy Rule, *Sharing the Wind*, THE ENVIRONMENTAL FORUM, Vol. 27, No. 5, pp. 30–33 (September/October 2010), and Troy A. Rule, *A Downwind View of the Cathedral: Using Rule Four to Allocate Wind Rights*, 46 SAN DIEGO LAW REV. 207 (2009).

2 It is worth noting that additional setback distances are often required for turbines near homes or public rights-of-way and justifiable to the extent they are needed to protect others from turbine noise or physical dangers.

3 *See* Kimberly Diamond & Ellen J. Crivella, *Wind Turbine Wakes, Wake Effect Impacts, and Wind Leases: Using Solar Access Laws as the Model for Capitalizing on Wind Rights During the Evolution of Wind Policy Standards*, 22 DUKE ENVTL. L. & POL'Y F. 195, 202 (2011).

4 According to one source, an increase in the long-term mean wind speed at a turbine site from six meters per second to 10 meters per second can increase a turbine's energy productivity by 134 percent. *See The Importance of the Wind Resource*, Wind Energy: The Facts, available at www.wind-energy-the-facts.org/the-importance-of-the-wind-resource.html (last visited February 24, 2014).

5 *See, e.g.*, S. Lee, et al., ATMOSPHERIC AND WAKE TURBULENCE IMPACTS ON WIND TURBINE FATIGUE LOADING, p. 12 (National Renewable Energy Laboratory, December 2011), available at www.nrel.gov/docs/fy12osti/53567.pdf (last visited June 19, 2013) (noting that "downstream" turbines in a study yielded higher "damage equivalent loads," suggesting that the "turbulent wakes from upstream turbines can have a significant impact" in increasing turbine fatigue for up to seven rotor diameters behind the upwind turbine). *See also* Kenneth Thomsen & Poul Sorensen, *Fatigue Loads for Wind Turbines Operating in Wakes*, 80 J. WIND ENGINEERING & INDUSTRIAL AERODYNAMICS 121 (1998) (noting that turbine "fatigue" or wearing of turbine parts increased by between 5 percent and 15 percent from wakes of upwind turbines for one offshore wind farm in Denmark).

6 BLACK'S LAW DICTIONARY 453 (4th ed. 1968). The full maxim is *cujus est solum,*

ejus est usque ad coelum et ad inferos. Please refer to Chapter 1 for a more detailed description of the *ad coelum* doctrine and its history.

7 *See* Charles C. Read & Daniel Lynch, *The Fight for Downstream Wind Flow*, LAW 360 (May 25, 2011), available at www.law360.com/articles/247122/ the-fight-for-downstream-wind-flow (last visited June 21, 2013).

8 *See* Lauren Donovan, *Two Energy Projects Competing for the Wind*, BISMARCK TRIBUNE (Feb. 22, 2008), available at http://bismarcktribune.com/news/local/ two-energy-projects-competing-for-the-wind/article_4bd1f0d6-6616-512b-970f-b4301800f774.html (last visited June 21, 2013).

9 Id.

10 Id.

11 Read & Lynch, *supra* note 7.

12 *See* id.

13 *See* id.

14 *See* id.

15 *Muscarello v. Winnebago County Board*, 702 F.3d 909, 912 (7th Cir. 2012).

16 Id.

17 Id.

18 Id. at 427.

19 *See Muscarello v. Ogle County Bd. of Com'rs*, 131 S.Ct. 1045 (2011).

20 *See Muscarello*, *supra* note 15 at 911.

21 WYO. STAT. ANN. § 34-27-103(a) (2011).

22 *See, e.g.*, Global Wind Energy Council, *Global Wind Report: Annual Market Update 2012* at 8 (2013), available at www.gwec.net/wp-content/ uploads/2012/06/Annual_report_2012_LowRes.pdf (last visited June 24, 2013) (noting that China was second only to the United States in new wind energy generating capacity in 2012 and leads Asia in that category).

23 *See* Marc Howe, *Chinese Regional Government Claims Wind Energy Is 'State-Owned'*, WINDPOWER MONTHLY (June 19, 2012), available at www. windpowermonthly.com/article/1136930/chinese-regional-government-claims-wind-energy-state-owned (last visited June 24, 2013).

24 *See, e.g.*, MONT. CODE ANN. § 70-17-404 (2011); NEB. REV. STAT. § 76-3004 (2009); N.D. CENT. CODE § 17-04-04 (2012); S.D. CODIFIED LAWS § 43-13-19 (2004); WYO. STAT. ANN. § 34-27-103(b) (2011).

25 *See Contra Costa Water Dist. v. Vaquero Farms, Inc.*, 58 Cal. App. 4th 883 (Cal. Ct. App. 1997).

26 For a more detailed discussion of wind rights severance and new statutes clarifying whether it is enforceable, *see* Troy Rule, *Property Rights and Modern Energy*, 20 GEO. MASON L. REV. 803, 811–12 (2013).

27 OTSEGO COUNTY ORDINANCES § 18.5.3.3 (2003).

28 *See* Order Establishing General Wind Permit Standards, Docket No. E, G-999/ M-07-1102, 7–8 (Minn. Pub.Util. Comm'n, Jan. 11, 2008), available at http://mn.gov/commerce/energyfacilities/documents/19302/PUC%20Order%20 Standards%20and%20Setbacks.pdf (last visited June 20, 2013).

29 *See* id.

30 *See* Donovan, *supra* note 8 (noting that a spokesman for FPL involved in a proposed wind energy project in rural North Dakota "wondered if the entire project would be feasible" if the local government imposed five-rotor-diameter setbacks on it).

31 Id. at 4.

32 Such policies requiring general payments to host communities to promote greater local acceptance of wind energy are prevalent in parts of Europe. *See, e.g.*, Sally Bakewell, *U.K. to Share More Wind-Power Wealth with Local Residents*,

Bloomberg (June 6, 2013), available at www.bloomberg.com/news/2013-06-05/u-k-to-share-more-wind-power-wealth-with-local-residents.html (last visited June 6, 2013) (describing regulatory changes that expanded the U.K.'s long standing policy of compensating host communities in connection with wind farms).

33 *See* North Dakota Engrossed House Bill No. 1460 (2011).

34 Id. at § 1.1.

35 *See* Rebecca Beitsch, *Bill Requires Wind Turbine Compensation Sharing*, BISMARCK TRIBUNE (Feb. 7, 2011), available at http://bismarcktribune.com/news/local/govt-and-politics/2011_session/bill-requires-wind-turbine-compensation-sharing/article_f3a7baaa-3277-11e0-82c9-001cc4c002e0.html (last visited June 24, 2012) (quoting a wind energy lobbyist who opposed North Dakota's H.B. 1460 on the ground that requiring payments to neighbors under a compulsory unitization statute for wind energy could drive up development costs).

36 *See Muscarello, supra* note 15 at 915.

37 *See* D. L. Elliot, et al., WIND ENERGY RESOURCE ATLAS OF THE UNITED STATES, ch. 2, pp. 3–4, National Renewable Energy Laboratory (1987) (noting, for instance, that, in the Columbia River corridor, "terrain variations cause considerable local variability in the wind resource").

38 For a more complete analysis of wind rights allocation schemes and the option-based policy approach described in this chapter, *see generally* Troy A. Rule, *A Downwind View of the Cathedral: Using Rule Four to Allocate Wind Rights*, 46 SAN DIEGO LAW REV. 207 (2009).

4 Renewable energy vs. wildlife conservation

Humans are not the only neighbors of wind and solar energy projects. Millions of species of animal life also inhabit this planet, and they too can be affected when giant wind turbines, solar power plants, or other forms of renewable energy development sprout up nearby. Renewable energy projects displace, injure, or kill countless birds, bats, and other animals each year. Wind turbines can slice off eagle wings, disrupt prairie chicken mating rituals, and hemorrhage the lungs of rare bats. Solar mirrors can confuse migratory birds in search of water holes[1] and can even concentrate sunlight and mortally burn raptors in midair.[2]

Given the wide range of threats that renewable energy development can pose to animals, it is no surprise that some conservationists view renewables as hazards to precious wildlife populations. Over the years, concerns over wildlife impacts have delayed or halted scores of proposed renewable energy projects throughout the world. Developers have encountered opposition based on their projects' potential effects on wild birds, bats, bears,[3] tortoises,[4] lizards,[5] rats,[6] toads,[7] foxes,[8] porpoises,[9] and even bighorn sheep.[10]

Ironically, renewable energy development might be one of the most important ways of promoting the long-term preservation of the earth's vast array of wildlife. By reducing demand for fossil fuel-generated power, renewables indirectly decrease global carbon emissions that could be contributing to global warming—an effect that could ultimately have far more catastrophic consequences for a much wider range of animals.[11] In that sense, evidence that a proposed renewable energy project would directly harm certain animals does not necessarily justify preventing its construction or operation. Of course, the mere fact that renewables might broadly benefit wildlife should not give developers license to trample over or endanger critical species at a particular project site, either.[12] How much attention, then, should policymakers and developers give to impacts on wildlife in the siting and design of renewable energy facilities?

A cost–benefit analysis for renewables and wildlife?

Some environmental ethicists might argue that renewable energy developers have an absolute moral obligation to protect animal life, regardless of the potential benefits of renewable power.[13] To them, evidence that a proposed wind or solar energy project would likely disrupt the habitat of even a single bald eagle or desert tortoise is sufficient cause to halt the project.

The reality is that nearly all forms of real estate development—including energy development—adversely impact wildlife to some degree, and the existence of these adverse impacts does not necessarily mean that construction should come to a halt. Construction of a single building often displaces or kills thousands of common ants, but no one argues against new construction on that account because there are probably millions of similar ants in the same region. Such a building is thus unlikely to threaten the continued existence of any particular ant variety or have any problematic impact on local ant population numbers.

Of course, the analysis quickly becomes more complicated when a wind or solar energy project threatens to adversely affect an animal species with much lower population numbers. Human disruptions of the habitats of these rarer creatures can implicate far greater social costs and thus cannot be as quickly dismissed as the inevitable consequences of social progress.

On the other hand, to preserve every living creature from the impacts of energy consumption, humankind would have to cease using energy altogether. Even if great strides in energy efficiency are made in the coming decades, the basic functioning of the global economy will always require some baseline amount of electricity generation. To the extent that this energy demand is not met through wind turbines, solar panels, and other renewables, it will necessarily come from sources such as nuclear energy, coal, and natural gas that can harm animal populations in other ways.

In other words, unless the world is willing to regress to the pre-industrial age, an energy strategy that spares every living creature from harm is not a viable option. As inhumane as it may seem, there is some socially optimal number of raptors that should die annually from collisions with wind turbine blades, and that number is greater than zero. Faced with this unpleasant reality, policymakers have no reasonable choice but to search for some efficient balance between renewable energy development and wildlife conservation, recognizing the need for compromise on both sides of this conflict. Despite what some ethicists might like to believe, a weighing of costs and benefits is the only practical approach to these issues.

Weighing uncertain benefits and uncertain costs

Unfortunately, cost–benefit analyses of conflicts between renewables and local wildlife are anything but easy. The benefits side of such equations requires some guesstimate of the aggregate societal benefit of clean renewable power. Kilowatt-for-kilowatt, renewable electricity generation is almost certainly less harmful to the planet and its wildlife population as a whole than the coal- or gas-fired generation that it most typically displaces. But how much less harmful are renewables?

Not surprisingly, opinions differ widely regarding the relative benefits to wildlife of renewable energy development over conventional fossil fuel-dependent energy generation. On the one hand, coal and oil production and combustion clearly adversely impact the environment.[14] Petroleum development and consumption can adversely affect animals through spills at wells or pipelines, toxic air emissions, contamination of storm water runoff, and other means. Coal-based energy strategies can likewise impose severe harm on wildlife in multiple ways. For example, the open cooling system at a single coal-fired power plant in the U.S. state of Massachusetts has been blamed for killing about 16 billion fish eggs and larvae.[15] And generating electricity from coal extracted at surface mines uses far more land area per kilowatt hour than generating electricity on commercial wind energy farms, potentially disrupting more habitats in the process.[16]

On the other hand, some wildlife advocates might take the view that a given wind farm's threat to a specific, endangered animal species is far more troubling than the generalized wildlife hazards associated with fossil fuel-generated energy. Even though wind farms are responsible for less than 1 percent of human-caused bird deaths each year, it may nonetheless be worthwhile to challenge a particular renewable energy project because of its potential impacts on a single animal.[17] For instance, in light of the tremendous investment that governments have made to save California condors from extinction, one could easily argue the death of such a bird would impose a greater social loss than the deaths of several dozen common seagulls or fish.

Unfortunately, it is far easier to understand this principle in the abstract than to apply it in practice. Would a given wind farm generate a net benefit to society or to the natural environment if it inadvertently took the life of one California condor but spared 1,000 seagulls from fatal harms due to additional offshore oil extraction? What if only 500 seagulls would be spared? Or just 100 seagulls? Even if policymakers had perfect information about the types and quantities of species that a given renewable energy project would indirectly spare over its existence, reaching any sort of policy consensus based on such data would be extremely challenging.

Likewise, estimating the *harm* that a given renewable energy project will cause to wildlife during its years of operation can be equally difficult. Renewable energy projects can injure animals directly through electrocution, collisions, or other direct contact, but they can also cause harms in less obvious ways, disrupting critical habitat areas, interfering with important survival activities such as migration or reproduction, or affecting creatures or plants[18] that are critical to an animal's food chain.[19] In addition, some of the environmental costs of renewable energy development occur at a great distance from the development site. The mere extraction of minerals used in renewable energy devices can harm animal habitats, and the manufacturing of turbines or solar panels requires energy that may be generated in wildlife-harming ways. Hazardous substances in photovoltaic solar cells and wind turbines can also pose risks to animals if they are improperly disposed of at the end of a project's life cycle.

Even if all of the harms and benefits to specific animal species from a proposed renewable energy project were somehow perfectly known, the analysis would not stop there. Other questions would inevitably remain. Some species are capable of generating tourism dollars, pollinating crops, and even helping to control pests—those benefits would have to be measured, too.[20] Other species might unexpectedly hold cures for diseases or keys to innovations with tremendous social value. An accurate harm estimate would need to incorporate these and still other factors.

In summary, uncertainty and ambiguity inevitably complicate attempts to accurately balance renewable energy development with competing interests in wildlife protection. Courts increasingly seem to recognize the tension between these two policy interests,[21] and even conservation groups have struggled to reconcile the complex relationship between them. Over the years, conservation groups have been somewhat divided in their support for or opposition to wind energy in areas where protected wildlife is present.[22] That said, many prominent groups such as the Audubon Society, American Bird Conservancy, and Royal Society for the Protection of Birds have each recently expressed support for wind energy development so long as developers assess wildlife impacts prior to construction and follow government guidelines for siting and design. This apparent trend suggests that wildlife advocates' opposition to renewable energy development may gradually diminish as the industry matures.[23]

In the meantime, the best that stakeholders can do is to improve their understanding of how renewable energy development can harm wildlife populations and to be careful when crafting policies to govern these conflicts. Fortunately, it is possible to minimize most of these impacts by assessing a given project's potential threats to wildlife early in the planning process and implementing all reasonable and cost-effective measures for mitigating significant threats.

Assessing a project's threat to wildlife

When a proposed renewable energy project poses some sort of threat to an animal species, initial questions about the potential conflict generally seek to assess how serious the harm will be. These inquiries typically require permit applicants or other stakeholders to gather information about the extent to which the species is already endangered and the likely severity of the project's impact on its population.

Identifying species of concern

Federal laws in the United States protect a long list of specifically identified species.[24] The Endangered Species Act (ESA)[25] makes it unlawful to "take" any threatened or endangered species on that list.[26] Among other things, to "take" an animal under the ESA means to "harass, harm …, wound, kill …, or collect" it.[27] Renewable energy facilities that harm, harass, wound or kill protected animals can certainly fall within the scope of that broad definition.[28] Hefty fines and as much as a year of jail time can attach to violations of the statute.[29]

Under the ESA, developers of real estate projects of any kind that may adversely affect any threatened or endangered species must first consult with the U.S. Fish and Wildlife Service (FWS), the government agency primarily responsible for administering the statute. Species listed as "endangered" are those that the FWS deems to be "in danger of extinction in all or a significant portion" of their range.[30] Species that are "likely to become" endangered "in the foreseeable future" are designated as "threatened" and also qualify for protection.[31]

Unfortunately, renewable energy developers cannot rely solely on published species lists or straightforward guidelines to determine their ESA compliance obligations in connection with a potential renewable energy project. Even species that are not presently listed as endangered or threatened may be "species of concern" entitled to protection under the ESA in some circumstances.[32] Complicating matters even further is the fact that the ESA does not even encompass the full set of wildlife protection laws potentially implicated by a renewable energy project. Multiple separate protection statutes exist in the United States for certain types of birds as well.[33]

Outside of the United States, many nations have similar wildlife protection laws. For example, the European Union has enacted significant endangered species protection legislation comparable to the ESA.[34] Renewable energy developers in all corners of the world must therefore be careful to adequately educate themselves about the regulatory regime and requirements applicable in their jurisdictions, even when their projects would seem to have no obvious impacts on endangered species.

Estimating likely impacts on species of concern

If it is determined that a proposed renewable energy project is likely to harm a species of concern, the debate then often shifts to questions about the severity of the harm. Numerous factors can influence the severity of a renewable energy project's impact on an important animal population. For instance, when a threatened animal population is very small in number, even minor habitat or food cycle disruptions can have material impacts so more stringent restrictions may be warranted. Likewise, animals with long life spans that mature late and have low reproduction rates are particularly vulnerable to development impacts because it can take longer for their populations to rebound.[35] Even an animal's physical traits can make it more or less susceptible to danger from a given renewable energy project. For example, wind turbines can be particularly hazardous to certain bat species that seem to be unusually attracted to spinning wind turbine blades.[36]

Strategies for mitigating renewable energy's impacts on wildlife

When there is significant risk that a proposed renewable energy project could disrupt a vulnerable wildlife species, government officials are likely to seek modifications to the development plan aimed at avoiding or mitigating those impacts. Such changes can take a wide range of forms, from minor design tweaks to total relocation of the project.

Onsite mitigation measures

Relatively minor schematic changes can often allow developers to reduce the potential adverse impacts of a renewable energy project on habitats and the surrounding natural environment. Over the years, researchers have identified several such onsite mitigation strategies. Some of these strategies are relatively easy and inexpensive to implement and are even promoted within the wind energy industry itself. However, some other onsite mitigation strategies can greatly diminish a project's profitability and even doom a project altogether. Negotiations between government siting officials and developers often involve some discussion of both types of onsite mitigation measures.

Some wildlife impact mitigation strategies in the context of renewable energy development involve siting and design elements implemented at the time of construction, while other strategies require ongoing implementation throughout the life span of a renewable energy project. For wind energy projects, this latter category of mitigation measures includes such things as turning off turbines during certain bird or bat migration periods.[37] Unlike design-based mitigation requirements, these ongoing mitigation efforts cannot be enforced solely at a project's final completion inspection. Instead, they require at least periodic monitoring, which can make them relatively more costly in practice.

Offsite mitigation measures

Increasingly, governments are demanding that developers make offsite mitigation commitments in connection with renewable energy projects as well. These commitments generally require developers to somehow offset certain wildlife impacts of their projects through activities away from the project site.

One common example of this type of mitigation measure is known as a habitat mitigation lease. Under this type of agreement, a renewable energy developer leases a large, privately-owned area that provides a similar habitat for the animal species threatened by the developer's project. Under the terms of the lease, the landowner agrees to abstain from developing the leased area for at least the life span of the renewable energy project. Such leases help to offset the adverse habitat impacts of the renewable energy projects by preserving comparable habitat elsewhere.

When properly implemented, these sorts of offsite mitigation measures can be a valuable and efficient means of balancing the competing goals of wildlife conservation and development of the world's most productive renewable energy sites. As described in previous chapters, in some regions the quality of wind energy resources can vary significantly by location. When wildlife species are adaptable such that habitat preservation activities are less location-sensitive than renewable energy resources, offsite habitat mitigation strategies can allow for the use of precious energy resources without compromising wildlife conservation goals in the process.

Of course, the threats to wildlife and other potential social costs of renewable energy development in some locations are so serious that allowing the project to proceed cannot be justified, even if the developer were to take all reasonable mitigation measures. In these situations, the best course of action may be to not build any project on the site.

As wind energy development has spread across the globe in recent decades, its wildlife conflicts that have drawn the most attention have involved birds and bats. Because opposition relating to these creatures is likely to continue to affect wind energy siting decisions well into the future, each receives special attention in this chapter. The penultimate section of the chapter then describes conflicts between solar energy development and the desert tortoise.

Wind energy vs. birds

Arguably the most well-known type of conflict involving wildlife and renewable energy is the tension between wind energy development and birds. Wind farms are estimated to kill hundreds of thousands of birds across the world each year.[38] At times, wildlife advocates have sharply criticized these sorts of clashes, describing them as an "outrage" and an "ecological disaster."[39] The most graphic bird deaths result from direct collisions with massive wind turbine blades. However, "collector" transmission lines within wind energy projects can also electrocute some large birds, whose long wingspans can allow them to simultaneously contact multiple wires.[40] And turbine noise and other less direct habitat disruptions are sometimes capable of reducing bird populations near wind energy projects as much as or more than collision deaths.[41]

What species of birds are the most vulnerable to harms from wind farms? How severely do wind energy projects really harm bird populations? And what strategies are available to help minimize these impacts? The following paragraphs respond to these questions and highlight some ways that bird impact issues have been addressed in the wind energy context.

Bird species commonly implicated in wind energy projects

The list of bird varieties impacted by wind energy development is quite lengthy and continues to grow. Somewhat surprisingly, the majority of the avian species killed by most wind energy projects are passerines, better known as songbirds.[42] As a whole, songbirds are far more prevalent in most areas of the world than the large, majestic birds that are often the focus of wind farm opposition groups. Because their collective numbers are so large, greater quantities of these small birds have had lethal encounters with wind energy facilities.

Of course, wind farms are better known for their threats to eagles and other raptors. These stately creatures can be attracted to breezy ridges and hilly areas where wind resources are particularly favorable for wind energy development.[43] Sadly, wind turbines can be deadly hazards to these animals. As one author describes it, "the vortices created by blade tips revolving at up to 200mph can destabilize such large birds, plunging them into a fatal collision."[44] A recent study concluded that at least 67 golden and bald eagles were killed by wind farms in the United States from 2008 to 2012.[45]

In Europe, wind farms have adversely affected populations of white-tailed eagles, which are that continent's largest birds of prey.[46] According to one recent study, the birds may actually be attracted to the "rotor-swept zone" where the turbine blades rotate, making it particularly susceptible to collisions and potentially helping to explain the relatively high number of deaths to the species in the vicinity of wind farms.[47]

In the United States, developers have had to relocate or alter numerous proposed wind energy projects in efforts to protect eagles. The bald eagle is the United States' national bird and has long enjoyed strong federal statutory protections, along with the golden eagle. In addition to the ESA's generic protections for these birds, two United States statutes offer further protection for eagles. The United States' Bald and Golden Eagle Protection Act (BGEPA) prohibits the "tak[ing]" of bald and golden eagles.[48] As was similarly mentioned above, to "take" a bird under FWS's Wind Energy Guidelines[49] includes to "disturb", which means to "agitate or bother ... to a degree that causes, or is likely to cause ... (1) injury to an eagle, (2) a decrease in its productivity ..., or (3) nest abandonment, by substantially interfering with normal breeding, feeding, or sheltering behavior."[50] Wind energy development that harms an eagle or its prey or habitat can certainly fall within the scope of such language. Convicted violators of the BGEPA can receive fines of up to $5,000 or sentences of up to one year in prison.[51]

Eagles in the United States are additionally protected under theMigratory Bird Treaty Act (MBTA),[52] which provides protections for them and more than 1,000 other birds.[53] Migratory birds are birds that relocate across great distances on a seasonal basis, often to breed or to winter in warmer climates. Although some studies have suggested that many migratory birds tend to fly high above wind farms during migration periods,[54] large numbers of these birds still die at wind farms. California's notorious Altamont Pass project exemplifies how destructive to bird populations wind farms can be when they are sited within bird migration routes. According to one source, more than 2,700 birds die from that single project each year, including more than 1,100 raptors.[55]

The MBTA, which was initially enacted in 1918 and has been described by the FWS as the "cornerstone of migratory bird conservation protection in the United States,"[56] implements the nation's international treaties with Canada, Japan, Mexico, and Russia to protect birds that migrate between these countries. Its provisions also prohibit the "tak[ing]" of birds on its list of protected species and provide for criminal penalties for violators.[57]

Interestingly, concerns about multiple bird species that rarely fly as high as a commercial wind turbine blade have also triggered strong opposition to wind farms on occasion. For example, in mountain regions of the western United States, government officials have delayed or rejected several wind farm proposals over fears about habitat impacts for the sage grouse.[58] The sage grouse was a candidate for endangered species listing in the United States as of early 2013.[59] Sage grouse and similar species "exhibit extreme avoidance of vertical features," making these birds particularly vulnerable to habitat fragmentation and other disruptions from wind farms.[60] Evidence suggests that the presence of towering wind turbines can deter members of the closely-related "prairie chicken" species found in the central United States from engaging in critical mating rituals.[61]

In addition to migratory birds and grouse, countless other bird species have been known to complicate the wind farm development approval process. Conservationists in the United States have raised concerns about wind energy's impacts on the endangered whooping crane, interior least tern, and piping plover in recent years, prompting the wind energy industry to help fund research to protect these species.[62] Fears of impacts on the highly-endangered California condor are also beginning to create new obstacles for wind energy development.[63] And the number of turbines in the first phase of the massive London Array offshore wind energy project in the UK was reduced from 258 to 175 to protect red-throated diver habitats in the Thames Estuary.[64] Given the rapid growth of wind energy throughout the world, the list of birds affected by wind farms will only expand in the coming years.

Estimating a proposed wind farm's potential impacts on birds

Because of the hazards that wind farms can sometimes create for statutorily protected bird species, cooperation with government wildlife agencies is an essential task item in any wind energy development plan. Within the United States, the FWS is the primary federal regulatory body that administers wildlife protection statutes relating to onshore wind farms. According to the agency's published guidelines, wind energy developers should initiate discussions with the FWS regarding possible "species of concern" during its preliminary site evaluation of any project site.[65] That initial contact with the FWS typically commences a back-and-forth site evaluation process specifically designed to identify the project's potential threats to wildlife, assess the severity of such threats, and formulate plans for addressing them. This "suggested communications protocol" requires a developer to assess such things as the project's potential to create habitat fragmentation and the presence of plant species and "congregation areas" critical to the animal's habitat.[66]

To assist in the information-gathering process, the FWS or similar agencies may even require that biologists make "reconnaissance level site visit[s]" to gather and confirm data on a project's potential habitat impacts.[67] Among other things, biologists may conduct "point counts" or other observational studies to help them better estimate the prevalence of birds within a proposed project area.[68] They may also search for raptor nests in the vicinity of proposed turbines or transmission facilities.[69]

Mitigating bird impacts in wind energy development

Once the information-gathering process is complete and wildlife agency officials have a solid grasp of the wildlife risks associated with a proposed wind energy project, agency officials next focus their attention on identifying how a developer might reduce those risks through various mitigation strategies.

The most obvious mitigation strategy for wildlife protection in the wind energy context is the siting of the project itself. Critical habitat areas of highly endangered birds such as the California condor may be wholly unfit for wind or solar energy projects, let alone any development. No amount of offsite or onsite mitigation may be adequate to address the risks associated with development in these sorts of highly sensitive areas. Fortunately, such high-risk areas are relatively few and far between. In most situations, renewable energy development can subsist with wildlife through careful pre-construction planning and routine monitoring of projects throughout their years of operation.

At wind farm sites where risks to birds are manageable, developers can employ a wide range of practices to help to mitigate bird impacts. One common practice is to simply avoid sensitive locations within the project when micro-siting turbines and supporting infrastructure. For example, adjusting turbine rotor heights based on pre-construction bird density data can allow developers to avoid placing rotors in migration routes or at altitudes where most local birds tend to fly. The FWS or similar agencies may also require buffer areas around geological features known to attract raptors such as cliffs or prairie dog colonies. Or, they may impose similar buffers of up to a mile around raptor nests or critical grouse nesting areas.[70]

Incorporating certain design features into a wind energy project can also mitigate its hazards to birds. Among other things, developers can bury intra-project transmission lines or omit external ladders from turbine towers so as not to create unnecessary perching areas for birds.[71] In some cases, developers have tried painting turbine blades with colors intended to scare off birds, or emitting sound or radar waves that dissuade birds from approaching turbine blades.[72] Such strategies can be highly effective, so long as they themselves do not inadvertently lead to increased habitat fragmentation.[73]

Another possible design approach is to install radar systems that notify project operators of approaching bird flocks and allow for the slowing or shutting down of turbines while the birds pass through the area.[74] These sorts of technology-based solutions are particularly promising as ways to reduce conflicts between wind farms and wildlife. As they are perfected over time and become more commonplace, they have the potential to significantly reduce wind farm-related animal deaths in the coming decades.

Even post-construction, wind farm operators can use several different strategies to continue mitigating wildlife hazards associated with their projects. For example, operators can periodically remove carcasses near the bottoms of wind turbines to avoid attracting vultures to such areas. Likewise, they can change the types of agricultural crops grown beneath turbines to reduce the presence of birds that favor a specific crop.[75] Collecting dead birds from beneath turbines and tracking counts of bird kills over time can also help wind farm operators to detect trends in bird deaths and identify periods of the year when slowing or shutting down turbines may be warranted.

Bird deaths and "incidental take" permits

Even when developers apply the sorts of mitigation measures described above, some birds inevitably die from wind farms. According to a 2013 peer-reviewed study by the *Wildlife Society Bulletin*, wind farms kill more than 573,000 birds every year in the United States alone, including approximately 83,000 raptors.[76] This harsh reality can raise perplexing policy questions. If a developer of a wind farm has taken all government-mandated precautions to mitigate hazards to birds, should the developer still be subject to penalties when inadvertent bird deaths occur? If so, are there any circumstances in which developers should be excused from such liability?

In some countries, the price tag for even unintentional bird deaths can be incredibly high. In March of 2013, a wind farm in eastern Nevada in the United States drew media attention because it discovered a dead golden eagle on its project site. Even though the bird's death appeared to have been purely accidental and the wind farm quickly turned the dead eagle over to the FWS and reported the case, the wind farm developer reportedly faced fines of up to $200,000 for that single bird's death.[77] And in November of the same year, a different company in the United States agreed to pay $1 million in fines in connection with the deaths of 14 golden eagles and other birds at two wind farms in the state of Wyoming.[78]

In spite of these examples, governments can sometimes be reluctant to enforce bird protection laws against wind farms because they are fearful that such enforcement could deter wind energy development. In fact, wind energy developers can sometimes get permits that expressly protect them from liability for unintentional bird deaths. Developers of wind energy projects in the United States in areas where protected birds may be threatened can apply to obtain "incidental take" permits from the federal government.[79] These permits effectively give wind farm developers and operators a limited license to kill protected birds.[80] The FWS only issues incidental take permits if a laundry list of requirements are met, including the applicant's submittal of a habitat "conservation plan."[81] These plans typically describe mitigation measures the developer will take to limit its threat to protected birds. Obtaining such permits can be crucial for wind energy developers of privately owned lands where deaths to protected birds are likely to occur.

Not surprisingly, the very existence of such permits and of policies aimed at promoting renewable energy at the expense of wildlife preservation has inflamed some wildlife advocates. Animal rights activists and some media organizations in the United States have criticized the Obama administration for neglecting to prosecute dozens of eagle deaths at wind farms over the course of Obama's presidency.[82] In addition, the administration has drawn attention to the revision of federal regulations to allow wind farm developers to obtain permits to incidentally kill eagles for up to 30 years in connection with their projects—a much longer permit period than the previous five-year rule.[83] Because the administration has continued to prosecute eagle deaths at *oil*-related facilities, it has been accused of applying a double standard.[84] It remains an open question whether the broader benefits of renewable energy justify this purported double standard as it relates to wildlife protection. What does seem clear is that these sorts of debates will only become more common as wind energy plays an ever larger role in the global energy economy.

Wind energy vs. bats

Although conflicts between wind turbines and birds tend to garner more headlines, wind turbines kill roughly as many bats as they do birds. By one researcher's estimation, wind farms are responsible for more than 450,000 bat deaths each year.[85] Conflicts between wind energy development and bats are similar in many respects to those involving birds, but some important differences exist that are worthy of mention.

Bats' unique vulnerability to danger from wind farms

Bats are so elusive that most people know relatively little about them. Roughly one fifth of all mammal species on earth are bats, and bats live in nearly every corner of the globe.[86] Most bats residing in temperate climates eat only insects and play an important role in the food chain and in agriculture. According to one estimate, the natural pest control provided by insectivorous bats saves United States farmers more than $3.7 billion per year.[87] However, although numerous species of bats enjoy statutory protections under the European Union's Habitats Directive,[88] few of the bat species materially affected by wind energy development in the United States are on the U.S. endangered species list.[89]

Some bat species seem to be inexplicably attracted to wind turbines.[90] And even though bats make effective use of sound waves and echo location to navigate and have little difficulty avoiding turbine blades themselves, they can be defenseless against the major drops in air pressure behind rotating turbine blades.[91] Research suggests that the majority of bat deaths near wind farms are due to pulmonary barotrauma—fatal lung damage from exposure to rapid air pressure decreases near the blades of massive, spinning wind turbines.[92] Several mass killings of bats at wind farms over relatively short time periods have been attributed to this effect. In one documented case, between 1,500 and 4,000 bats died at a 44-turbine wind farm in a single year.[93]

A large proportion of bat fatalities at wind farms is believed to occur during seasonal migrations. Although few bat species engage in cross-continental migrations like those of many migratory birds, bats do annually migrate across modest distances to caves or other underground locations where they hibernate during the cold winter months.[94] Because they tend to travel in the dark of night, relatively little is known about bats' nocturnal migration patterns.[95] This lack of information complicates wind farm operators' efforts to predict when such migrations may occur and adjust accordingly. Bats also tend to have slower reproduction rates and longer life expectancies than birds, making it more difficult for their populations to rapidly recover from disruptions.[96] All of these factors make certain species of bats particularly sensitive to having wind farms nearby.

Mitigation strategies for protecting bats near wind farms

In light of the aforementioned challenges, what can wind energy developers and operators do to prevent massive killings of bats on their projects while still generating wind power? To date, researchers have been able to uncover relatively few answers to this question. At least one study suggests that painting wind turbines purple could be an onsite mitigation strategy capable of reducing fatalities to bats,[97] although few developers have actually tried it.

The most promising mitigation technique for protecting bats seems to be the practice of increasing turbine "cut-in speeds" during night hours when bats are most likely to be in the air. Typically, commercial wind turbines are designed and programmed to begin rotating and generating electricity when wind speeds reach eight to nine miles per hour.[98] A 2009 study co-sponsored in the United States by the Bats and Wind Energy Cooperative and the Pennsylvania Game Commission concluded that increasing the cut-in speed on one Pennsylvania wind farm's turbines to approximately 11 miles per hour resulted in dramatic reductions in bat fatalities.[99]

Unfortunately, many operators of wind farms are not likely to view the approach of increasing turbine cut-in speeds as an appealing solution to this problem. The reductions in bat deaths in the Pennsylvania study were likely attributable to the fact that, due to the increase in cut-in speeds, the turbines involved in the study rotated less frequently. In other words, the turbines were not generating power when wind speeds were between 8 and 11 miles per hour, even though they would have been generating power at such wind speeds under normal operations. The Pennsylvania study was conducted during relatively low-wind-speed months, so overall decreases in energy production from the technique were relatively modest.[100] However, any strategy that requires a major reduction in energy generation is costly to wind farms and thus bound to be less popular with the wind energy industry.

A case study of clashes between bats and wind farms: Animal Welfare Institute v. Beech Ridge Energy LLC

The 2009 case of *Animal Welfare Institute v. Beech Ridge Energy LLC*[101] illustrates the inherent tension between protecting bats and promoting wind energy. The defendants in the case, Beech Ridge Energy LLC and Invenergy Wind LLC, were wind energy developers of a wind farm in Greenbrier County, West Virginia, in the United States. Multiple plaintiffs filed suit against the developers to enjoin operation of the project on the ground that the project would otherwise result in the incidental "taking" of numerous Indiana bats—a species of bat appearing on the endangered species list and thus protected under the ESA.

The court in *Beech Ridge* clearly recognized that the dispute at hand was a mere microcosm of the much broader conflict between wildlife conservation and the renewable energy movement. To quote the court opinion, "This is a case about bats, wind turbines, and two federal policies, one favoring protection of endangered species and the other encouraging development of renewable energy resources."[102]

Ultimately, the court upheld an injunction against operation of the wind farm except during summer months when bats would not be at risk. The court seemed to take the view that Congress intended to place protecting endangered species above facilitating renewable energy development. Even though the defendant's wind farm would have supplied clean, renewable power to more than 50,000 homes,[103] the court found that Congress had "unequivocally stated that endangered species must be afforded the highest priority."[104] Accordingly, the court proved unwilling to engage in any sort of weighing of costs and benefits between renewable energy and wildlife protection, seemingly deferring to the judgment of Congress on such policy questions.

After the court released its decision, the developer of the Beech Ridge wind energy project began work on a permit application for an incidental take permit for the second phase of its wind energy project. If granted, the permit would shield the developer from liability for the incidental deaths of any Indiana bats or Virginia big-eared bats resulting from that phase of the wind farm.[105] The developer's habitat conservation plan, required as part of its incidental take permit application, included an offer to help fund offsite conservation efforts for the two protected bat species potentially impacted by the project.[106]

The *Beech Ridge* case sent a clear message to renewable energy developers who may be tempted to develop projects without fully cooperating with the FWS. At least within the United States, courts are unlikely to give serious consideration to arguments that the social and environmental benefits of renewable energy justify ignoring FWS requirements. Any future pardons or special exceptions from such compliance for renewable energy developers are likely to have to originate in legislative bodies, not in the courts.

Solar energy vs. the desert tortoise

In addition to harming birds and bats, renewable energy development has the potential to adversely impact several other forms of wildlife. Many of these are desert creatures that inhabit arid climates where there are strong renewable energy resources and few residents capable of being disturbed by a large wind or solar energy project.

In recent years, there has been growing interest in siting solar energy facilities in remote desert areas. Some of these new projects are concentrating solar power (CSP) plants, which involve the use of hundreds or thousands of mirrors to concentrate sunlight toward a central tower where it heats water and drives a steam turbine. Other desert solar energy projects are simply massive, ground-mounted PV solar arrays.

Siting wind and solar energy farms in sparsely populated rural desert areas may reduce the likelihood of neighbor conflicts, but sometimes its disruptions of protected animal populations can be just as difficult to overcome. As the utility-scale solar energy development industry has grown in the past couple of decades, it has faced near-constant resistance because of its potential harms to a diverse list of desert creatures that includes everything from squirrels and foxes to lizards and toads.

In the southwestern United States—a region with superb solar resources and aggressive government support for renewable energy—conservationists concerned about impacts on the desert tortoise have been a major impediment to solar energy development. Because its name appears on the U.S. threatened species list,[107] the desert tortoise receives significant protection under the ESA. Particularly in the states of California and Nevada, this listing has complicated efforts to develop solar energy plants on vast public lands aimed at harvesting more of the region's superb solar energy resources.

Several characteristics of desert tortoises make them particularly susceptible to population declines from the introduction of solar energy projects in their habitat areas. These slow-moving reptile creatures typically do not begin reproducing until they are about 12 years old, and only 2 percent of hatchlings born in the wild survive.[108] They eat mostly cacti and other desert plants that solar energy developers may partially clear in connection with their projects, and solar mirror installations have the potential to penetrate and disturb their buried eggs or underground burrows.[109]

Events surrounding the Ivanpah CSP project in California's Mojave Desert exemplify the conflict between desert tortoise habitat conservation and solar energy development in this part of the world. Initially, U.S. government officials estimated that the massive and innovative Ivanpah project would impact less than 40 tortoises. The developer, BrightSource Energy, obtained an incidental take permit under the ESA in connection with the project, agreeing among other things to help fund the safe relocation of tortoises it displaced. Based on those estimates and the developer's habitat conservation plan, the Sierra Club and similar conservation groups supported the project, concluding that this project's climate change-fighting benefits exceeded its localized wildlife habitat risks.[110]

Unfortunately, as construction on the Ivanpah project proceeded, it became increasingly apparent that many more tortoises were being affected than originally expected. By June of 2012, construction activities had displaced at least 144 desert tortoises, prompting federal officials to delay further construction to allow the gathering of better information about tortoise population impacts at the site.[111] A conservation group law suit followed, aimed at enjoining any further construction.[112]

As of mid-2013, construction on the Ivanpah project had resumed and tortoise relocation efforts had proven quite successful. However, BrightSource's mitigation efforts came at a hefty price: according to one source, the developer spent more than $56 million "caring for and relocating" tortoises associated with the project.[113] And Ivanpah is just one of several solar energy projects in the southwestern United States that has suffered resistance based on desert tortoise concerns.[114] Hopefully, developers will be able to develop better and less costly mitigation techniques as solar energy development continues in that region in the coming years.

Balancing wildlife conservation and renewable energy in the next century

There is no simple solution for the inherent tension between renewable energy development and wildlife conservation. Real estate development of nearly every kind has always impacted local wildlife populations and likely always will. At the root of almost all of these conflicts is a basic weighing of social priorities. Should policymakers do any- and everything possible to protect a single identified animal species when it faces near-certain population declines from a new renewable energy project within its habitat areas? Or, should policymakers allow such population declines when they are necessary to increase the world's renewable energy generating capacity—a cause aimed at ultimately protecting humans and a much broader range of animal species?

The optimal balance between encouraging renewable energy and protecting wildlife lies somewhere in the middle of these two competing interests and differs for every type of wildlife and every renewable energy project. Research, innovation, information sharing, and candid public discussion about this difficult balance will be crucial as humankind continues to seek it throughout the remainder of this century and beyond.

Notes

1 *See* U.S. Bureau of Land Management and U.S. Dept. of Energy, *Summary of Public Scoping Comments Received During the Scoping Period for the Solar Energy Development Programmatic Environmental Impact Statement* at 7 (Oct. 2008), available at http://solareis.anl.gov/documents/docs/scoping_summary_report_solar_peis_final.pdf (last visited June 10, 2013).

2 *See generally* M. D. McCrary, et al., *Avian Mortality at a Solar Energy Power Plant*, JOURNAL OF FIELD ORNITHOLOGY, Vol. 57, No. 2, pp. 135–41 (1986).

3 Significant controversy regarding potential impacts on bear populations has surrounded a proposed commercial wind energy project in Vermont, USA, in recent years. For an academic perspective on this conflict, *see generally* Reed Elizabeth Loder, *Breath of Life: Ethical Wind Power and Wildlife*, 10 VT. J. ENVTL. L. 507 (2009).

4 Concerns about possible impacts on desert tortoise habitats complicated the permit approval process for multiple proposed wind farm projects in California

and Nevada, USA, in recent years. *See, e.g.*, Hadassah M. Reimer & Sandra A. Snodgrass, *Tortoises, Bats, and Birds, Oh My: Protected Species Implications for Renewable Energy Projects*, 46 IDAHO L. REV. 545, 560–61 (2010) (describing opposition from environmental groups to proposed Searchlight Wind Project in Southern Nevada based in possible disruption of tortoise habitat); U.S. Bureau of Land Management, *McCoy Solar Energy Project* (2013), available at www.blm.gov/pgdata/etc/medialib/blm/wo/MINERALS__ REALTY__AND_RESOURCE_PROTECTION_/energy/priority_projects. Par.56888.File.dat/McCoy%20Solar%20fact%20sheet.pdf (describing offsite mitigation measures in connection with McCoy Wind Project in California aimed at protecting desert tortoise habitat).

5 *See, e.g.*, McCoy Solar Energy Project, *supra* note 4 (describing developer's purchase of mitigation lands to protect habitat of Mojave fringe-toed lizard).

6 Specifically, fears over impacts on giant kangaroo rat populations have generated opposition to utility-scale solar energy projects in California in recent years. *See, e.g.*, David Sneed, *Giant Kangaroo Rat Puts Kink in California Valley Solar Project*, THE TRIBUNE (San Luis Obispo, CA) (Sept. 11, 2010), available at www.sanluisobispo.com/2010/09/11/1284985/giant-kangaroo-rat-puts-kink-in.html (last visited June 10, 2013) (describing how the discovery of burrowing holes of endangered giant kangaroo rats that a proposed solar energy project would likely require a scaling-back of the project).

7 In particular, conflicts with the Amargosa toad in Nevada, USA, have complicated the siting of solar energy projects. *See* Reimer & Snodgrass, *supra* note 4 at 576–77.

8 Solar energy projects have been blamed for the deadly disruption of desert kit fox habitats in California, USA. *See, e.g.*, Louis Sahagun, *Canine Distemper in Kit Foxes Spreads in Mojave Desert*, LOS ANGELES TIMES (Apr. 18, 2012), available at http://articles.latimes.com/2012/apr/18/local/la-rme-0418-foxes-distemper-20120418 (last visited June 10, 2013) (describing a possible link between stress from "hazing techniques" used to force the foxes from their habitats in connection with the Genesis Solar Energy Project in southern California may have contributed to the outbreak of a deadly disease among the local population of this endangered mammal species).

9 Climate Wire, *Offshore Wind Turbines Keep Growing in Size*, SCIENTIFIC AMERICAN (Sept. 19, 2011), available at www.scientificamerican.com/article. cfm?id=offshore-wind-turbines-keep (last visited Oct. 11, 2013) (describing how environmental groups were "lobbying for construction to be halted" on a wind farm off the shore of Germany because they were "worried about the effect of construction noises on porpoises").

10 One recent effort to mitigate impacts of a wind energy project on bighorn sheep populations involved the Ocotillo Wind Energy Facility in California, USA. To review the mitigation and monitoring plan for the project, visit www. ocotilloeccmp.com/Wild1s_PBS_MMP.pdf (last visited June 10, 2013).

11 *See IPCC Technical Paper V: Climate Change and Biodiversity* at 16 (§ 6.21) (Apr. 2002), available at www.ipcc.ch/pdf/technical-papers/climate-changes-biodiversity-en.pdf (last visited June 13, 2012) (describing evidence that climate change may "increase species losses").

12 For one prominent scholar's recent examination of the unique tension between green energy development and wildlife conservation, *see generally* J. B. Ruhl, *Harmonizing Commercial Wind Power and the Endangered Species Act through Administrative Reform*, 65 V. AND. L. REV. 1769 (2012).

13 For a discussion of the ethical, as opposed to economic, arguments against renewable energy development in valued habitat areas, *see generally* Lodor,

supra note 3 at 522–26 (noting that some might characterize utilitarian analyses of conflicts between wildlife and renewable energy development like those above as "anthropocentric" or excessively focused on human interests above those of animals and arguing that such approaches have "undergone steady attack for upwards of twenty-five years").

14 *See, e.g.*, E&P Forum/UNEP, *Environmental Management in Oil & Gas Exploration & Production* at 17–20 (1997), available at www.ogp.org.uk/pubs/254.pdf (last visited June 12, 2013) (showing a table of potential adverse environmental impacts of oil and gas exploration and production). For similar information about the impacts of coal exploration and production, visit the United States Environmental Protection Agency's webpage on the subject at www.epa.gov/cleanenergy/energy-and-you/affect/coal.html (last visited June 13, 2013).

15 *See* Alexa Burt Engelman, *Against the Wind: Conflict over Wind Energy*, 41 ENVTL. L. REP. NEWS & ANALYSIS 10549, 10551 (2011).

16 *See* id. at 10533 (noting that "over a 30-year period, a surface coal mine will use 21, 844 acres of land, while an average wind array will use 4,720 acres to produce the same amount of power").

17 *See* Avi Brisman, *The Aesthetics of Wind Energy Systems*, 13 N.Y.U. ENVTL. L.J. 1, 72 (2005) (citing research by the American Wind Energy Association).

18 Although this chapter focuses primarily on wild animal life, plant species certainly can suffer harms near renewable energy projects as well. *See, e.g.*, Jeffrey E. Lovich & Joshua R. Ennen, *Wildlife Conservation and Solar Energy Development in the Desert Southwest United States*, 61 BIOSCIENCE 985 (2011) (noting that dust emissions from the installation of solar energy facilities in desert climates can "adversely influence the gas exchange, photosynthesis, and water usage of Mojave Desert shrubs" and "can physically damage plant species through root exposure, burial, and abrasions to their leaves and stems") (citation omitted).

19 In the context of wind energy development, the U.S. Fish and Wildlife Service has more specifically delineated a list of potential wildlife risks. *See* U.S. Fish and Wildlife Service, *Land-Based Wind Energy Guidelines*, at vi (2012), available at www.fws.gov/windenergy/docs/weg_final.pdf (last visited June 12, 2013).

20 For a more detailed analysis of these issues, *see generally* Derek Bertsch, *When Good Intentions Collide: Seeking a Solution to Disputes between Alternative Energy Development and the Endangered Species Act*, 14 SUSTAINABLE DEV. L.J. 74, 92–94 (2011).

21 At least one scholar has made mention of this fact. *See* Alexandra B. Klass, *Energy and Animals: A History of Conflict*, SAN DIEGO J. CLIMATE & ENERGY L. 159, 190 (2011–12) (noting that "courts are very aware of the tensions that exist between the [U.S.] Endangered Species Act … on the one hand […] and federal and state policies promoting renewable energy … on the other").

22 *See* id. at 204 (arguing that, "[w]hile environmental groups have in the past been fairly uniform in their skepticism or outright opposition to many aspects of traditional energy development …, renewable energy elicits a much more mixed response").

23 *See* Engelman, *supra* note 15 at 10553 (noting support from the Audubon Society and American Bird Consevancy). To review the RSPB's official policy on wind energy development, visit the group's website at www.rspb.org.uk/ourwork/policy/windfarms/ (last visited June 12, 2013).

24 To search for endangered species by state within the United States, visit the Endangered Species webpage of the U.S. Fish and Wildlife Service website at www.fws.gov/endangered/ (last visited June 12, 2013).

25 Endangered Species Act of 1973, 16 U.S.C. §§ 1531-1544 (2006).

26 16 U.S.C. § 1538(a)(1)(B).

27 16 U.S.C. §1532(19).

28 For the definitions of "harass" and "harm" under the ESA, *see* 50 C.F.R. 17.3.

29 *See* 16 U.S.C. § 1540(b).

30 16 U.S.C. § 1532(6).

31 16 U.S.C. § 1532(20).

32 FWS guidelines define "species of concern" to include any bird that "has been shown to be significantly adversely affected by wind energy development" and "is determined to be possibly affected" by a given project. Fish and Wildlife Service Land-Based Wind Energy Guidelines, *supra* note 19 at 63.

33 More detailed descriptions of the United States' Migratory Bird Treaty Act, and Bald and Golden Eagle Protection Act, which are examples of these other wildlife protection statutes, are provided later in this chapter. *See generally* pages 81–82, *infra*.

34 For a primer on endangered species legislation in Europe, *see generally* Ian F. G. McLean, *The Role of Legislation in Conserving Europe's Endangered Species*, 13 CONSERVATION BIOLOGY 996 (1999).

35 *See generally* Mark Desholm, *Avian Sensitivity to Mortality: Prioritising Migratory Bird Species for Assessment at Proposed Wind Farms*, 90 JOURNAL OF ENVIRONMENTAL MANAGEMENT 2672 (2009) (noting that many waterfowl and birds of prey are particularly vulnerable to population declines from collisions with wind turbines because of these species tend to live longer and lay fewer eggs than many other bird varieties).

36 *See* Umair Irfan, *Bats and Birds Face Serious Threats from Growth of Wind Energy*, NEW YORK TIMES (August 8, 2011), available at www.nytimes.com/cwire/2011/08/08/08climatewire-bats-and-birds-face-serious-threats-from-gro-10511.html?pagewanted=all (last visited June 11, 2013).

37 See Ryan Tracy, *Wildlife Slows Wind Power*, WALL STREET JOURNAL (Dec. 10, 2011), available at http://online.wsj.com/article/SB10001424052970203501304577088593307132850.html (last visited June 11, 2013).

38 Dan Frosch, *A Struggle to Balance Wind Energy with Wildlife*, NEW YORK TIMES A18 (Dec. 17, 2013), available at www.nytimes.com/2013/12/17/science/earth/a-struggle-to-balance-wind-energy-with-wildlife.html?_r=0 (last visited Dec. 17, 2013).

39 *See* Christopher Booker, *Wind Turbines: "Eco-friendly"—But Not to Eagles*, THE TELEGRAPH (March 13, 2010), available at www.telegraph.co.uk/comment/columnists/christopherbooker/7437040/Eco-friendly-but-not-to-eagles.html (last visited June 12, 2013).

40 *See* AMERICAN WIND ENERGY SITING HANDBOOK § 5.1.1 (2008), available at www.awca.org/sitinghandbook/downloads/awca_siting_handbook_feb2008.pdf (last visited June 23, 2013).

41 *See, e.g.*, *Wind Farms Cause Decline in Bird Population—RSPB*, THE TELEGRAPH (Sept. 26, 2009), available at www.telegraph.co.uk/earth/earthnews/6231580/Wind-farms-cause-decline-in-bird-population-RSPB.html (last visited June 13, 2013) (citing a study by the Royal Society for the Protection of Birds and Scottish Natural Heritage primarily attributing declines near 12 wind farms in upland regions of the United Kingdom to the "noise and development" itself and that "strikes with turbines" were "less likely" to have been the cause).

42 *See* AMERICAN WIND ENERGY SITING HANDBOOK, supra note 40 at § 5.1.1.

43 Booker, *supra* note 39.

44 Id.

45 See Dina Capiello, *Study Finds Wind Farms Kill Eagles*, BOSTON GLOBE (Sept. 12, 2013).

46 *See Wind Farm "Hits Eagle Numbers"*, BBC NEWS (June 23, 2006), available at http://news.bbc.co.uk/2/hi/5108666.stm (last visited June 13, 2013).

47 *See generally* Espen Lie Dahl, et al., *White-tailed Eagles (Haliaeetus Albicilla) at the Smøla Wind-power Plant, Central Norway, Lack Behavioral Flight Responses to Wind Turbines*, 37 WILDLIFE SOCIETY BULLETIN 66 (2013).

48 16 U.S.C. § 668(a).

49 *See* U.S. Fish and Wildlife Service Land-Based Wind Energy Guidelines, *supra* note 19 at 2.

50 50 C.F.R. 22.3.

51 16 U.S.C. § 668(a).

52 16 U.S.C. § 703 et seq.

53 For the list of protected species under the MBTA, *see generally* 50 C.F.R. 10.13.

54 *See, e.g.*, Guyonne Janss, *Bird Behavior In and Near a Wind Farm at Tarifa, Spain: Management Considerations*, National Renewable Energy Laboratory Wind-Wildlife Impacts Literature Database, Doc. 1173 at 112 (2000), available at http://nrelpubs.nrel.gov/Webtop/ws/avianlt/www/web_data/Record;jsessioni d=54428F622B6C4F7111335F2CC5691584?m=1101 (last visited Dec. 17, 2013) (describing research finding that several bird species flew at higher altitudes in response to the presence of wind farms).

55 *See* Engelman, *supra* note 15 at 10551.

56 Fish and Wildlife Service Land-Based Wind Energy Guidelines, *supra* note 19 at 2.

57 16 U.S.C. §§ 703 & 707.

58 Examples of such delayed or abandoned projects include the Simpson Ridge Project in Carbon County, Wyoming and the China Mountain project near Twin Falls, Idaho. *See* Scott Streater, *Wind Power Industry Retreating from Wyo., Citing Sage Grouse Concerns*, NEW YORK TIMES (Aug. 7, 2009), available at www.nytimes.com/gwire/2009/08/07/07greenwire-wind-power-industry-retreating-from-wyo-citing-93484.html?pagewanted=all (last visited June 12, 2013).

59 For more information on the sage grouse and its candidacy for endangered species listing, visit the U.S. Fish and Wildlife Service website at www.fws.gov/mountain-prairie/species/birds/sagegrouse/ (last visited June 13, 2013).

60 *See* James M. McElfish Jr. & Sara Gersen, *Local Standards for Wind Power Siting: A Look at Model Ordinances*, 41 ENVTL. L. REP. NEWS & ANALYSIS 10825, 10833 (2011).

61 *See* Jim Efstathiou Jr., *Prairie Chicken Mating Dance Threatens Texas Projects (Update 1)*, BLOOMBERG (Aug. 26, 2009), available at www.bloomberg.com/apps/news?pid=newsarchive&sid=aldysneqgVeA (last visited June 10, 2013).

62 For information about the Great Plains Wind Energy Habitat Conservation Plan involving these species and the lesser prairie chicken, visit the Plan's website at www.greatplainswindhcp.org/ (last visited June 12, 2013).

63 *See* Felicity Barringer, *Turbine Plans Unnerve Fans of Condors in California*, NEW YORK TIMES at A17 (May 25, 2013).

64 *See* Edward Platt, *The London Array: The World's Largest Offshore Wind Farm*, THE TELEGRAPH (July 28, 2012), available at www.telegraph.co.uk/earth/energy/windpower/9427156/The-London-Array-the-worlds-largest-offshore-wind-farm.html (last visited Oct. 15, 2013).

65 Fish and Wildlife Service Land-Based Wind Energy Guidelines, *supra* note 19 at 5.

66 Id.
67 Id.
68 Id. at 28.
69 *See* id. at 29.
70 *See* id.
71 Many of these sorts of mitigation measures for wind energy installations are set forth in the U.S. Bureau of Land Management's "programmatic environmental impact statement," which applies to wind energy projects on public lands governed by that federal agency. *See* Bureau of Land Management, Final Programmatic Environmental Impact Statement on Wind Energy Development on BLM-Administered Lands in the Western United States, 5-65 and 5-66 (2005), available at http://windeis.anl.gov/documents/fpeis/maintext/Vol1/Vol1Ch5.pdf (last visited June 11, 2013).
72 *See* Bertsch, *supra* note 20 at 84.
73 *See* Irfan, *supra* note 36.
74 This strategy has recently been highlighted as a way of reducing risks to California condors. *See, e.g.*, Louis Sahagun, *Feds Won't Prosecute Wind Farm if Turbine Blades Kill a Condor*, Los Angeles Times (May 24, 2013), available at www.latimes.com/news/science/sciencenow/la-sci-sn-california-condor-wind-farm-20130524,0,3874393.story (last visited June 12, 2013).
75 *See* Janss, *supra* note 54 at 113.
76 *See* K. Shawn Smallwood, *Comparing Bird and Bat Fatality-rate Estimates Among North American Wind-energy projects*, 13 Wildlife Society Bulletin 19 (March 2013).
77 See Henry Brean, *Eagle Death at Nevada Wind Farm Brings Federal Scrutiny*, Las Vegas Review-Journal (March 25, 2013), available at www.reviewjournal.com/business/energy/eagle-death-nevada-wind-farm-brings-federal-scrutiny (last visited June 12, 2013).
78 *See* Emma G. Fitzsimmons, *Wind Energy Company to Pay $1 Million in Bird Deaths*, New York Times A23 (Nov. 24, 2013), available at
79 *See* 16 U.S.C. § 1539(a)(1)(B).
80 Such permits, commonly referred to as "Section 10 Incidental Take Permits" are not issuable on projects having any nexus to federal action. For a more detailed description of this limitation on incidental take permits under U.S. federal law, *see generally* Hadassah M. Reimer & Snodgrass, *supra* note 4 at 549–51.
81 16 U.S.C. § 1539(a)(2).
82 *See* Dina Capiello, *Wind Farms Get Pass on Deaths of Eagles, Other Protected Birds*, Seattle Times (May 15, 2013), available at http://seattletimes.com/html/localnews/2020993836_windfarmsbirdsxml.html (last visited June 12, 2013).
83 *See* Frosch, *supra* note 38.
84 *See* id.
85 *See* Paul M. Cryan, *Wind Turbines as Landscape Impediments to the Migratory Connectivity of Bats*, 41 Envtl. L. 355, 364 (2011).
86 *See* Cryan, *supra* note 85 at 357 (noting that "there are approximately 5400 species of mammals on Earth, of which about 1100 are bats" and that "[b]ats occur nearly everywhere but Antarctica and some remote islands").
87 *See* Justin G. Boyles, et al., *Economic Importance of Bats in Agriculture*, 332 Science 41 (Apr. 1, 2011).
88 To view a full list of bat species protected under the EU Habitats Directive, visit the Finnish Environmental Institute's website at www.ymparisto.fi/default.asp?node=12219&lan=en#a1 (last visited June 12, 2013).
89 *See* Cryan, *supra* note 85 at 367.

90 *See* AMERICAN WIND ENERGY ASSOCIATION SITING HANDBOOK, *supra* note 40 at § 5.1.2.1. Interestingly, roughly one half of all bats killed by wind farms in the United States are hoary bats. Cryan, *supra* note 85 at 364.

91 *See* Irfan, *supra* note 36.

92 *See generally* Erin Baerwald, et al., *Barotrauma Is a Significant Cause of Bat Fatalities at Wind Turbines*, 18 CURRENT BIOLOGY 695 (2008), cited in Reimer & Snodgrass, *supra* note 4 at 564.

93 *See* Justin Blum, *Researchers Alarmed by Bat Deaths from Wind Turbines*, WASHINGTON POST (Jan. 1, 2005), available at www.washingtonpost.com/wp-dyn/articles/A39941-2004Dec31.html (last visited June 12, 2013).

94 *See* Cryan, *supra* note 85 at 358.

95 *See* id. at 359–60.

96 Many species of bats have only one offspring per year, and bats often live for ten years or more. *See* id. at 360–61.

97 For a basic discussion of this idea, *see* Laura Roberts, *Wind Turbines Should be Painted Purple to Deter Bats, Scientists Claim*, THE TELEGRAPH (Oct. 15, 2010), available at www.telegraph.co.uk/earth/earthnews/8066012/Wind-turbines-should-be-painted-purple-to-deter-bats-scientists-claim.html (last visited June 11, 2013).

98 *See Slight Change in Wind Turbine Speed Significantly Reduces Bat Mortality*, SCIENCE NEWS (Nov. 3, 2010), available at www.sciencedaily.com/releases/2010/11/101101115619.htm (last visited June 12, 2013).

99 *See* id. To review the full study, *see generally* Edward B. Arnett, et al., *Altering Turbine Speed Reduces Bat Mortality at Wind-Energy Facilities*, 9 FRONTIERS IN ECOLOGY AND THE ENVIRONMENT 209 (2011).

100 *See supra* note 98.

101 675 F. Supp. 2d 540 (D. Md. 2009).

102 Id. at 542.

103 *See* id. at 548–49.

104 Id. at 581.

105 To download Beech Ridge Energy LLC's incidental take permit application, visit www.fws.gov/westvirginiafieldoffice/beech_ridge_wind_power.html (last visited June 12, 2013).

106 *See Beech Ridge Wind Energy Project Habitat Conservation Plan* at 95, available on the FWS website at www.fws.gov/westvirginiafieldoffice/PDF/beechridgehcp/Beech%20Ridge%20Draft%20HCP%20(May%202012).pdf (last visited June 12, 2013).

107 For details about the desert tortoise and its ESA listing, visit the FWS website at http://ecos.fws.gov/speciesProfile/profile/speciesProfile.action?spcode=C04L (last visited June 13, 2013).

108 *See* Ken Wells, *Tortoises Manhandled for Solar Splits Environmentalists*, BLOOMBERG (Sept. 20, 2012), available at www.bloomberg.com/news/2012-09-20/tortoises-manhandled-for-solar-splits-environmentalists.html   (last visited June 13, 2013).

109 *See* id.

110 *See* Bertsch, *supra* note 20 at 93 (noting that the Sierra Club and the Nature Conservancy were supportive of the Ivanpah project despite its potential harms to the threatened desert tortoise).

111 *See* Wells, *supra* note 107.

112 *See* id.

113 *See* id.

114 *See* Reimer & Snodgrass, *supra* note 4 at 573–74 (describing desert tortoise-related issues for proposed solar energy projects in California's Broadwell Dry Lake area and at the proposed Beacon Solar project).

5 Solar energy vs. shade

Sunlight has always been critical to survival on this planet. The vast majority of the earth's living organisms rely on light and warmth from the sun for their continued subsistence. Plants that store solar energy through photosynthesis form the basis for food chains that sustain and nourish billions of humans and other creatures throughout the globe. Even the energy in the fossil fuels that heat our homes, propel our automobiles, and illuminate our computer screens originates from the sun.

On the other hand, concerns about global warming are a reminder of the potential problems of having too much solar radiation. Most scientists now believe that fossil fuel emissions are trapping excess solar heat in the earth's atmosphere and that this greenhouse effect could cause average temperatures to rise significantly and hazardously impact communities and ecosystems throughout the world.

Ironically, humankind is increasingly looking to sunlight for ways to combat global warming. Technologies that convert solar resources directly into electric power can offset some of the world's consumption of coal, oil, and natural gas and thereby help to reduce global emissions of greenhouse gases. As the demand for solar energy systems rises, economies of scale are enabling manufacturers to make these systems ever more efficient and cost-effective.[1] Consequently, for the first time in history, solar energy is beginning to have a noticeable impact in energy markets throughout the world.

Although the pace of utility-scale solar energy is growing, most solar energy development is "distributed" in nature, consisting of countless rooftop solar PV arrays spread throughout cities and towns. Distributed solar energy development is advantageous in that it does not require the installation of costly transmission superhighways through remote rural areas. On the other hand, the growth of distributed solar energy also means that landowners are increasingly installing solar energy systems close to homes and other types of development, thereby increasing the potential for conflicts.

As the global solar energy industry has blossomed in the past decade, so has interest in the controversial topic of "solar access." Shade is the arch nemesis of solar PV systems, rendering them far less productive than when they have direct exposure to sunlight. Even relatively minor shading can have major impacts on a solar array's electricity production. According to one study, shading just 10 percent of some solar panels can cause a 50 percent reduction in their electricity output.[2]

Although landowners can prevent trees or buildings on their own parcels from shading their solar energy systems, they often have far less control over trees or construction on *neighbors'* land. This risk of unwelcome shading from a neighbor's tree or new building can unnecessarily deter solar panel installations in jurisdictions without policies to adequately govern solar access conflicts.

[handwritten margin note: can it prevent neighbors from shading]

What is the solar access problem?

A simple example is helpful in illustrating how solar access conflicts can arise. Suppose that a landowner who is interested in solar energy (Solar User) plans to install an array of PV solar panels on the rooftop of her home. Although prices for PV panels have fallen dramatically in the past decade, solar arrays are still quite expensive. Accordingly, based on some preliminary research, Solar User predicts that it will take roughly 15–20 years for her to recoup the cost of the array through savings on her electricity bill. Solar User is more than willing to make that investment, so she purchases and installs the system.

Unfortunately, a couple of years after Solar User installs her rooftop solar array, she notices that a tree on her neighbor's property is beginning to shade the panels in the winter months during certain hours of the day. The neighbor's tree continues to grow and to shade an ever larger proportion of the solar array and for longer periods throughout the year, as shown in Figure 5.1 below. Not surprisingly, Solar User is beginning to notice reductions in the quantity of electric power her solar panels are generating and is seeing a consequent rise in her monthly electricity bills. When Solar User asks her neighbor to trim the tree to reduce its impact on her solar array, the neighbor refuses. Solar User feels helpless, seemingly unable to avoid the economic losses resulting from the shade from her neighbor's tree.

[handwritten margin note: risk]

This risk—that a neighbor could grow a tree or erect a new building that casts unwanted shade across the property line—is a factor that landowners must consider when contemplating whether to get a rooftop solar energy system. In some cases, parties in Solar User's position can convince neighbors to voluntarily sell solar access easements or covenants to protect against this risk. In fact, laws in many U.S. states specifically provide for the enforceability of solar access easements.[3] These easements prohibit servient landowners from occupying those portions of the airspace above their land with trees or structures that would shade the dominant owner's solar array. Unfortunately, some landowners are unwilling to grant these easements, even for a reasonable price. When a neighbor refuses to grant a solar access

[handwritten margin note: solution: solar access easements]

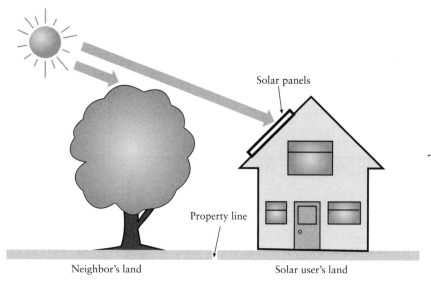

Figure 5.1 The solar access problem

easement, parties in Solar User's position may conclude that the risk of future shading is too great and thus elect not to install solar energy systems on their rooftop at all.

At first blush, an effective solution to the solar access problem just described would be to enact laws that prohibit landowners from growing trees or erecting structures on their land that shade solar energy systems on nearby parcels. However, formulating rules to govern solar access conflicts is far more complicated than that because keeping airspace open solely to prevent solar panel shading may not always be the most productive use of the airspace involved. As described briefly in Chapter 1, trees sequester carbon dioxide out of the atmosphere and can substantially reduce a building's need for energy-intensive air conditioning. And dense vertical construction that occupies airspace with tall buildings is one way of combatting urban sprawl and its energy excesses.

Should landowners be prohibited from erecting structures or growing trees on their property that shade a neighbor's solar panels? Or should they be liable for damages when such shading occurs? Or should landowners be free to occupy the airspace above their property within ordinary height restrictions, regardless of whether they shade nearby solar energy systems? These questions are similar to other questions posed in this book and are emblematic of the tough property rights trade-offs that complicate the sustainability movement.

The best solar access policies are those that incentivize landowners to keep airspace open only when preventing solar panel shading is a higher valued use of the space than occupying that space with buildings or trees. The challenge for policymakers is to craft workable laws that actually create those incentives for landowners in their jurisdictions. This chapter recounts the history of laws involving sunlight and analyzes several modern solar access policies developed after the advent of solar energy technologies. The chapter ultimately advocates a particular policy approach designed to promote optimal use of the scarce airspace resources implicated in disputes over solar access.

Solar access laws prior to the solar energy era

Although disputes over solar panel shading are relatively new, conflicts over access to sunlight have been around for centuries. A brief recounting of the history of laws relating to sunlight helps to put modern solar access issues into context.[4]

Direct sunlight has played a prominent role in indoor lighting throughout most of human history. Until the twentieth century, illumination from candles, lamps, and other fire-lit devices was the only alternative to natural sunlight. Because daytime sunlight was free of cost and created no smoke or fire risk, it was an important light source in many buildings. It is easy to understand the frustration that landowners in such settings must have felt when new buildings or walls erected on nearby parcels suddenly shaded the windows of their dwellings.

The ancient Romans recognized the tremendous importance of sunlight access in buildings and thus enacted laws to protect it. According to one pair of scholars, civil laws in ancient Rome required that builders either "leave [...] neighbors a minimum or reasonable amount of daylight" or obtain from neighbors a "servitude" allowing for a greater amount of light obstruction.[5] Although such laws surely offered some protection to landowners, ambiguity regarding what constituted a reasonable minimum quantity of daylight may have made such laws difficult to enforce.

The doctrine of ancient lights

Rights to sunlight access also enjoyed robust protections under English common law. Under the English "doctrine of ancient lights," a landowner who benefited from uninterrupted sunlight use on a parcel for a full prescriptive period could thereby obtain the equivalent of a perpetual easement protecting that sunlight access.[6] Dating back to at least 1610, this legal doctrine has existed in one form or another in the United Kingdom for over four centuries.[7] By the year 1832, 20 years was the clearly-established prescriptive period of uninterrupted use required to obtain rights to light under the rule.[8]

Although the doctrine of ancient lights still existed in the United Kingdom as of 2013,[9] few western courts outside that country ever

[handwritten margin note: Continuous use → perpetual easement]

embraced the doctrine. In early United States history, several state courts allowed landowners to obtain prescriptive easements for sunlight access.[10] However, by the turn of the twentieth century, courts in the U.S. were far less willing to enforce such rights.[11] Courts in Australia and New Zealand have also largely rejected the notion of prescriptive rights to light.[12]

doctrine being phased out (margin note)

Some courts and policymakers have opposed the doctrine of ancient lights because of its potential to preclude more dense development— a land use that often constitutes a more valuable use of scarce urban airspace.[13] Over time, modern electricity grids and the invention of the incandescent light bulb surely diminished support for the doctrine as well by making sunlight access far less critical to the enjoyment of land. Regardless, even in jurisdictions that embraced the doctrine, its mandatory twenty-year prescription period makes it of negligible value to most parties interested in protecting their investment in new rooftop solar arrays.

discourages dense development (margin note)

Nissho-ken: solar rights in Japan

Japan is another country with laws that offer some degree of solar access protection to landowners. In Japan, a nation with a relatively high population density and several cities filled with high-rise buildings, citizens have come to place a particularly high value on the benefits of direct sunlight access. Landowners in Japan enjoy some limited legal rights to sunlight access—something that Japanese courts have held is a critical element of healthy living conditions. This Japanese theory of rights to sunlight, or *nissho-ken,* generally prevents neighbors from developing land in ways that would deprive nearby parties of a specific minimum amount of sunlight access.[14]

right to sunlight access (margin note)

As Japan was rapidly rebuilding its nation after World War II, a rapid proliferation of dense development in the country led to hundreds of landowner disputes over sunlight access.[15] From 1968 to 1974, nearly 300 different citizens filed claims with the Tokyo District Court alone seeking injunctions against construction projects that threatened to interrupt the flow of sunlight and air to their homes.[16] In one such case, *Mitamura v. Suzuki,* the Japanese Supreme Court held that the plaintiff, Mitamura, was entitled to compensation for a new building's obstruction of sunlight to Mitamura's home.[17] Within a few short years after this 1972 decision, hundreds of Japanese cities had enacted laws aimed at protecting sunlight access for existing buildings.[18]

Unfortunately, although landowners in some parts of Japan possess some minimal solar access rights, it is not at all clear that the nation's *nissho-ken* laws are broad enough to protect against solar panel shading. Such laws seem to be aimed primarily at protecting enough sunlight and air to provide for a healthy life, not to protect an investment in a solar energy system.

not sure about implications for solar (margin note)

Sunlight access policies to promote solar energy development

Although laws in multiple countries have protected rights to sunlight at various times throughout history, access to sunlight soon took on new importance in the 1970s when solar PV energy systems first became widely available.[19] For the first time ever, landowners could install devices on their rooftops that converted sunlight into electric current capable of running everything from their television sets to their toasters. Coming upon the heels of the Arab oil embargo, these advancements in PV technologies generated a wave of interest in solar energy development in the United States.[20]

During this hopeful era for solar energy, some academicians and policymakers in the United States began to question whether current laws adequately protected rights to solar access. The substance of their inquiry was straightforward: were existing policies that provided little or no sunlight access protection still appropriate in an era of solar panels? Or were those laws outmoded, given the advent of solar energy technologies?

Rethinking solar access laws: Prah v. Maretti

In 1982, a state court in the United States drew attention for suggesting that recent innovations in solar energy technology might justify new legal rules that strengthened landowners' solar access rights. The plaintiff in the case of *Prah v. Maretti* was Glenn Prah, an owner of land in the state of Wisconsin. Prah had installed "solar collectors" on his home that supplied warm water and heat to the home for much of the year.[21] The defendant in the case, Richard Maretti, planned to construct a two-storey dwelling on an adjacent parcel situated immediately to the south of Prah's land. If constructed, Maretti's planned home would cast shade that substantially reduced the productivity of Prah's solar collectors.

When Prah learned of Maretti's construction plans, he contacted Maretti and requested that the new home be sited several feet back from the lot line to mitigate the shading problem.[22] When Maretti ignored Prah's requests and commenced building in the originally-planned location, Prah filed a legal claim seeking damages and an injunction to prevent construction from going forward. The lawsuit eventually reached the Wisconsin Supreme Court, where Prah argued that Maretti's proposed home would interfere with Prah's right to "unrestricted use of the sun and its solar power."[23]

In its published opinion, the majority in *Prah* noted that the increasing use of solar energy devices brought into question whether stronger solar access protections may be warranted under the law. In the majority's words:

> [A]ccess to sunlight has taken on a new significance in recent years. In this case the plaintiff seeks to protect access to sunlight, not for aesthetic reasons or as a source of illumination but as a source of energy. Access to sunlight as an energy source is of significance both to

the landowner who invests in solar collectors and to a society which has an interest in developing alternative sources of energy ... Courts should not implement obsolete policies that have lost their vigor over the course of the years.[24]

Concluding that it was conceivable for there to be a nuisance claim based on "unreasonable obstruction of access to sunlight" in the new solar energy age,[25] the court refused to grant summary judgment against Prah.

The dissenting justice in *Prah* took a very different view of the case. He pointed out that Maretti's proposed home complied with all applicable building and zoning laws, which did not require any consideration of shading of neighboring properties.[26] He also emphasized, as described in Chapter 1 of this book, that landowners hold property interests in the airspace above their land under the *ad coelum* doctrine and that forbidding Maretti to build the home within that airspace would violate his property rights.[27] Of course, these arguments ultimately failed, and the case was published suggesting that nuisance law may be broad enough to encompass the shading of a solar energy system in some circumstances.

The majority holding in *Prah* created a stir in legal and policy circles for its seeming deviation from longstanding U.S. law. However, few U.S. courts elected to follow *Prah*'s reasoning in the decades following the decision. Most remain unwilling to enforce solar access rights for solar energy users absent a written agreement or statutory rules requiring such enforcement.

Although United States' courts ultimately proved hesitant to adjust airspace rights to promote sunlight access for solar panels, state and local legislative bodies in the U.S. enacted a wide spectrum of solar access protection laws in the late 1970s and early 1980s. Many of these laws remain on the books today. Unfortunately, as the following subsections suggest, few of these statutes are very effective at governing solar access conflicts.

Solar access protection gone too far: California v. Bissett *and revisions to California's Solar Shade Control Act*

The state solar access statute that has garnered the most publicity in recent years is California's Solar Shade Control Act (SSCA).[28] The California state legislature enacted the SSCA in 1978, just two years after the state first began offering generous tax credits for solar energy projects.[29] Under the originally-enacted version of the SSCA, a landowner whose trees or vegetation substantially shaded a neighbor's qualifying solar energy system could be convicted of a public nuisance.[30] Those guilty of shading under that 1978 version of the statute could be ordered to raze their trees or trim their vegetation to stop the shading and could receive fines of up to $1,000 per day until they complied.[31]

By the mid-1980s, interest in solar energy had waned in the United States and California's SSCA had begun to gather dust. Courts cited its provisions only a small handful of times during its first quarter century on the books.[32] However, when interest in solar energy development finally re-blossomed in the early 2000s, the statute's aggressive public nuisance provisions quickly began to take on relevance again.[33]

A neighbor dispute arising in 2008 brought national attention to the SSCA. Richard Treanor and his wife, Carolynn Bissett, owned a home in a suburban neighborhood in Sunnyvale, California. The couple was undeniably sustainability-minded: they owned a Toyota Prius hybrid vehicle and took great pride in their backyard trees.[34] Treanor and Bissett had planted several redwood trees in their yard in the late 1990s, which Bissett said she hoped "to grow old with."[35] Unfortunately, the trees soon become the focal point of a bitter controversy when the couple's neighbor, Mark Vargas, decided to install solar panels on a trellis over the hot tub behind his home.

The parties to this dispute have given inconsistent accounts of what happened next. According to one newspaper article, Vargas claims that he politely asked Treanor and Bissett to "remove the trees and offered to pay for removal and replacement."[36] In contrast, Treanor and Bissett say that Vargas "brusquely" informed them of the law and told them that they had to cut back their trees.[37] Regardless, an ugly green-versus-green conflict between trees and solar panels was born. Treanor and Bissett refused to prune their trees, so the Santa Clara County District Attorney eventually filed a criminal public nuisance claim against the couple.[38]

After seven years of battling the issue before city officials and eventually in the courts, Treanor and Bissett were ultimately convicted of one count of a public nuisance under the SSCA for their refusal to trim their trees. Having already spent $37,000 in legal fees, this environmentalist couple painfully watched one of their trees get cut back to comply with the statute.[39] A barrage of national media covered the story, stirring lively debate over the California law behind it all.[40] Similar disputes had already begun popping up elsewhere in the state.[41]

Less than one year after the convictions of Treanor and Bissett, California's legislature amended the SSCA to remove those portions of the statute that had drawn the most criticism during the highly publicized case.[42] Under the statute's amended version, landowners can no longer be criminally convicted for public nuisance for shading a neighbor's solar energy system. Solar energy system owners now must bring *civil* suits against neighbors for shading under a private nuisance theory.[43] The state legislature's amendments also added exemptions for trees planted before a plaintiff's solar energy system was first installed.[44]

[margin handwritten note: now only civil nuisance suits]

Even after the amendments, the SSCA's provisions are arguably an unfair means of governing solar access conflicts. The statute is clearly a well-intended policy with a noble goal of promoting solar energy, but it still often disregards landowners' existing airspace rights. Neighbors of solar energy users generally hold property rights in the low-altitude airspace above their land under the *ad coelum* doctrine discussed in Chapter 1. In many situations, the SSCA creates the equivalent of uncompensated solar access easements through these neighbors' airspace, as roughly depicted in Figure 5.2. The statute imposes these easements on neighbors solely to benefit specific landowners with solar energy systems and provides no compensation for this loss of airspace rights.

The SSCA not only disregards existing property rights; it is also prone to inefficiency. The statute rests upon the tenuous presumption that protecting solar collectors from shading is a more socially valuable use of most suburban airspace than occupying that space with trees. As mentioned in Chapter 1, well-placed trees can offer lots of benefits in such settings, from reducing air conditioning and heating loads to improving stormwater drainage. Trees must be sited near homes and other buildings to provide these location-specific benefits. In contrast, many types of solar energy systems are far less location-dependent. They are capable of capturing roughly the same amount of solar radiation and generating equivalent amounts of energy regardless of where they are installed within a community.

As imperfect as the SSCA may be, some state solar access statutes are even more aggressive and more troublesome than California's law. For example, under the solar access statute enacted in the state of Massachusetts, landowners can unilaterally obtain "permits" that protect their systems from shading by neighbors without compensating neighbors

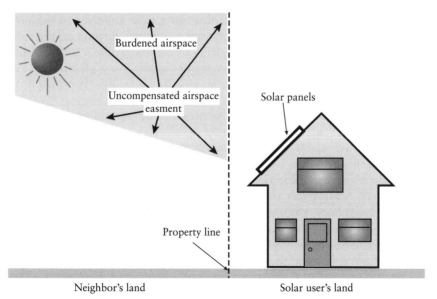

Figure 5.2 Airspace burdened under a solar access easement

for the consequent loss of their airspace rights.[45] Once landowners receive one of these permits, their neighbors are prohibited not only from growing trees that would shade the system but also from erecting buildings or other structures that would shade it. These policies generate efficient outcomes only in cases where keeping airspace open to protect solar access is more socially valuable than allowing trees or buildings within that space.

First-in-time rules for solar access and the fallacy of "solar rights"

Some solar access statutes frame these conflicts in a different way but are ultimately prone to creating the same sorts of suboptimal outcomes as the Massachusetts law just described. Among these are statutes that characterize solar access as a property right in and of itself, akin to water rights in some western U.S. states. The following article excerpt describes the shortcomings of these other, well-intended state laws.

> In New Mexico and Wyoming, a landowner can unilaterally acquire solar access rights across Neighbors' airspace, without compensating Neighbors, by being the first to make "beneficial use" of the airspace. A landowner who installs a qualifying solar collector, records a valid solar right instrument or declaration with the county clerk, and satisfies statutory neighbor notice requirements under these statutes acquires "solar rights." Solar rights acquired under these statutes are not rights in sunlight itself or in some other scarce resource for which private property rights did not previously exist. The New Mexico and Wyoming statutes define a "solar right" as a property right to an "unobstructed line-of-sight path from a solar collector to the sun" which "permits radiation from the sun to impinge directly on the solar collector." In essence, a solar right is an easement across a Neighbor's airspace for the specified purpose of solar access. New Mexico's statute even has language requiring that a solar right be "considered an easement appurtenant."
>
> In solar access disputes in New Mexico and Wyoming, "priority in time" supposedly "[has] the better right." Unfortunately, solar access conflicts are rarely disputes over competing solar access easements in which one Solar User erects a solar collector in the solar access path of another Solar User. Instead, such disputes are almost always between Solar Users seeking to obtain or enforce solar access rights and Neighbors with no interest in installing solar collectors who seek only to preserve existing airspace rights. In nearly every circumstance, Neighbors were "first-in-time" with respect to the Airspace Entitlement because they hold title to the surface estate directly below the airspace at issue. Although the New Mexico and Wyoming statutes are a well-intended effort to innovatively promote solar access, they ignore Neighbors' existing airspace rights and misapply the prior

race to obtain permit

appropriation doctrine. The statutes seem based on the presumption that neither Solar Users nor their Neighbors already possess rights in the airspace at issue. In truth, Neighbors of Solar Users do hold such rights under common law.

The New Mexico and Wyoming statutes are not the first-in-time rules they purport to be, but they do adjust or reallocate existing property rights among landowners based on priority in time of use. They can thus generate many of the same unintended consequences associated with first-in-time rules. California's Solar Shade Control Act and the Wisconsin and Massachusetts permit-based solar access statutes are like the New Mexico and Wyoming statutes in this regard. All of these statutes enable Solar Users to unilaterally acquire rights in or impose restrictions on Neighbors' airspace, but *only to the extent* that the airspace is not already occupied. Such approaches promote solar energy development by motivating Solar Users to install solar collectors quickly before Neighbors make use of the airspace needed for solar access. They may also, however, encourage opportunistic landowners to install solar panels with ulterior motives of acquiring a view easement across Neighbors' property or of preventing or delaying Neighbors' more productive uses. The rules might also motivate Neighbors to overdevelop their properties with trees or structures to avoid forfeiting their airspace rights to new Solar Users. Because they impose individualized burdens based on the needs of individual private landowners and without compensation, the rules are also more vulnerable to constitutional challenge.

(Troy Rule, *Shadows on the Cathedral: Solar Access Laws in a Different Light*, 2010 U. Ill. L. Rev. 851, 876–78 (2010))

Wyoming's and New Mexico's unusual solar access policies remain enforceable today. Only time will tell whether the same sorts of controversies that surrounded the SSCA will ultimately provoke legislators to modify these laws so that they are more efficient and more equitable to neighboring landowners.

Shade-based zoning, setbacks, and fences

Some solar access policies take a very different approach to those just described, seeking to prevent solar access conflicts ex ante through zoning and building codes rather than addressing solar access conflicts as they arise. Although this strategy certainly prevents solar access disputes, it too can lead to the under-use of scarce airspace resources. As the following excerpt argues, such laws are seldom desirable from a policy perspective.

[Another] approach to promoting solar access is to impose setbacks and height restrictions in zoning ordinances designed for the sole purpose of protecting solar access. For example, the City of Ashland, Oregon, has adopted a Solar Access Ordinance that imposes "[s]olar [s]etbacks" on nearly all private property within that jurisdiction. Under the ordinance, newly constructed structures must not cast a shadow taller than a certain height at the north property line. Greater structure heights are allowed at greater distances from northerly boundary lines. The City of Boulder, Colorado, has adopted similar provisions for certain zones within that city, referring to these height and setback requirements as "solar fence[s]". In both cases, the applicable height limitations vary among zones within the jurisdiction, allowing for greater building heights in certain commercial or industrial zones.

At first blush, shade-based setbacks are appealing in that they largely avoid the arbitrariness and strategic behavior problems that characterize some Lot-By-Lot Statutes. The rules apply uniformly within a given zone, like any ordinary setback or height restriction. They may also be less administratively burdensome than Lot-By-Lot approaches and may provide greater certainty to landowners.

Nevertheless, statutes that impose shade-based setbacks on all properties within an area—not just those situated near solar panels—are probably not cost justified. If the aggregate value of all the myriad airspace uses prohibited by a shade-based setback rule exceeds the aggregate value of solar access it preserves, the rule is likely inefficient. Less than 0.5% of residential rooftops in the United States presently hold solar panels. Thus, more than 99% of the time, Solar Users that hold the scarce Airspace Entitlement under these statutes do not exist. Until rooftop solar systems become far more prevalent, it seems a dubious presumption that the solar access protected under broad shade-based setbacks is of greater social benefit than the endless number of other legal uses of airspace within a jurisdiction that are prohibited by the setbacks.

(Troy Rule, *Shadows on the Cathedral: Solar Access Laws in a Different Light*, 2010 U. ILL. L. REV. 851, 878–80 (2010))

Given the shortcomings inherent in all of the above-mentioned policy approaches, it may be easy to understand why the majority of jurisdictions in the United States and across the planet have *no* laws to protect solar energy systems from shading. In most jurisdictions of the western world, landowners are generally not liable for shading their neighbors' solar panels, so solar energy users can only obtain protection against such shading through voluntarily negotiated private agreements. Of course, even this "hands off" policy approach seems unsatisfactory to those who favor solar energy because it wholly fails to address what can be a significant obstacle to distributed solar energy development.

The Iowa approach: better rules for solar access

Fortunately, there exists at least one other policy strategy to govern solar access that, if properly structured and implemented, allows a means of overcoming obstacles to solar energy development without disregarding existing airspace rights. The rural state of Iowa in the central United States appears to be the only known jurisdiction that has adopted a law resembling this alternative policy approach.

Iowa's solar access law was enacted in the early 1980s—roughly the same time period as most of the other solar access statutes applicable in the United States today.[46] In many ways, Iowa's law is similar to the Massachusetts law described above. It essentially provides that landowners who install qualifying solar energy systems on their properties can obtain solar access easements from neighbors to protect against shading. However, unlike those in Massachusetts, landowners in Iowa who seek such protection against shading must *purchase* it from neighbors at fair market value rather than receiving it for free.

Iowa's statute is relatively straightforward. Landowners in Iowa who plan to install solar energy systems on their land but are concerned about neighbor shading must first approach their neighbors and offer to purchase a solar access easement to protect against this shading risk. If a neighbor refuses to sell such an easement, landowners can apply to a local government entity for an order granting the easement.[47] Applicants who meet all of the statutory requirements for the easement get an option to purchase it from their neighbors.

Iowa's solar access easement application process

To obtain an order granting a solar access easement under Iowa's statute, applicants must satisfy several requirements aimed at minimizing local government involvement and the burdens placed on neighbors. Quoting directly from statute in Iowa Code § 564.A.4.1 (2013), each application must contain the following:

a. A statement of the need for the solar access easement by the owner of the dominant estate.
b. A legal description of the dominant and servient estates.
c. The name and address of the dominant and servient estate owners of record.
d. A description of the solar collector to be used.
e. The size and location of the collector, including heights, its orientation with respect to south, and its slope from the horizontal shown either by drawings or in words.
f. An explanation of how the applicant has done everything reasonable,

taking cost and efficiency into account, to design and locate the collector in a manner to minimize the impact on development of servient estates.

g. A legal description of the solar access easement which is sought and a drawing that is a spatial representation of the area of the servient estate burdened by the easement illustrating the degrees of the vertical and horizontal angles through which the easement extends over the burdened property and the points from which those angles are measured.

h. A statement that the applicant has attempted to voluntarily negotiate a solar access easement with the owner of the servient estate and has been unsuccessful in obtaining the easement voluntarily.

i. A statement that the space to be burdened by the solar access easement is not obstructed at the time of filing of the application by anything other than vegetation that would shade the solar collector.

The requirement in part *f.* above is particularly noteworthy. Part *f.* seeks to deter applicants from attempting to burden more neighboring airspace than is reasonably necessary to avoid solar panel shading. As will be described below, the requirement that applicants *purchase* their solar access easement rather than obtain it for free also helps to deter such behavior because the easement purchase price is likely to increase as the size of the burdened easement area grows.

Part *h.* of the application requirements is also significant in that it requires applicants to first try to negotiate the voluntary sale of a solar access easement before filing their application. This certification requirement encourages landowners to reach agreements to resolve solar access conflicts without government involvement when possible. Of course, the potentially greater involvement of expensive lawyers when applying for an order under the statute surely promotes voluntary resolution of these conflicts as well.[48]

Notice and hearing requirements

Once an applicant has filed a complete application under Iowa's statute, the local government's designated "solar access regulatory board" (which may in some cases be a local board of adjustment or other existing local government body) schedules a hearing and sends a notice to the neighbor whose airspace would be burdened.[49] The written notice, paid for by the applicant,[50] provides the date of the hearing and explains the neighbor's rights and how the process will proceed. At the hearing, the regulatory board reviews the applicant's materials and hears any objections from the neighbor. If the board is convinced that all statutory requirements are satisfied, it may issue an order for a solar access easement at that time.

Iowa's statute intelligently spells out some specific situations when boards are precluded from issuing an order for an easement, even though the application is complete. Like the provisions described above, these exceptions also help to prevent abuse of the statute and to promote more efficient use of the airspace involved. For instance, in recognition of the fact that trees can serve as valuable windbreaks or that their shade may reduce air conditioning use during hot months, a board may refuse to grant an easement order if it determines that "the easement would require the removal of trees that provide shade or a windbreak to a residence on the servient estate."[51] The statute also requires boards to deny applications in cases where the neighbor has "made a substantial financial commitment to build a structure that will shade the solar collector" within the previous six months.[52] This provision makes it more difficult for landowners to exploit the statute's provisions to hinder already-announced development projects.

Compensation to neighbors and recording of the easement

Prior to issuing an order for a solar access easement, a board operating under Iowa's solar access statute determines some amount of compensation due to the owner of airspace that would be burdened under the easement. Simple appraisal techniques are available to help determine this amount, which is to be the "difference between the fair market value of the property prior to and after granting the solar access easement."[53] Airspace is clearly more valuable in some areas than in others, so the appraised value of such easements can vary widely depending on the parties' location.

Once the easement's value is set, an applicant in Iowa may elect not to pay the compensation amount and to simply abandon the entire process. In some cases, that may be the most efficient outcome. For example, in urban downtown areas, market values for airspace may be quite high. In such situations, the required compensation amount for a solar access easement would presumably be high as well. Applicants for solar access easements in such locations are likely to be unwilling to pay such high amounts and thus will not purchase easements under the statute, allowing the airspace to be used for downtown buildings—a higher valued airspace use in that setting.

If an applicant elects to pay the set compensation amount, the board collects these funds and then issues a solar access easement that is recorded in the public records like any other easement.[54] The owner of the burdened neighboring property also receives the compensation funds.

In Iowa, solar access easements have no statutorily-set durational term—a missing element of what is otherwise a relatively well-written statute. However, when landowners with solar access easements obtained under Iowa's statute stop using their solar energy systems, neighbors can petition for removal of the easement.[55] This provision at least provides a means of eliminating a solar access easement that no longer is serving any productive purpose.

Merits of the Iowa approach

Iowa's approach of requiring landowners to pay fair value for involuntary solar access easements obtained under its solar access statute does more than just promote the fair treatment of neighbors and respect for their existing airspace rights. This requirement also encourages optimal use of the airspace involved in solar access conflicts. As discussed in Chapter 1, airspace is a scarce, valuable resource and is playing an increasingly important role in the sustainability movement. Solar access statutes like Iowa's that include compensation requirements make it possible for all landowners to obtain solar access protection, but incentivize them to do so only when that is the highest and best use of that airspace.

To illustrate these efficiency benefits of including a compensation requirement, consider again the solar access conflict in the *California v. Bissett* case described earlier in this chapter. That case received media attention because it pitted two "green" airspace uses against each other: trees and solar access protection. In many jurisdictions throughout the world, landowners like Mark Vargas who sought legal protection for their new solar arrays would be at the mercy of neighbors to voluntarily sell them that protection for a reasonable price. Vargas would have had to kindly ask Bissett and Treanor to sell him an easement. Given how fiercely the couple opposed the razing of their trees, Bissett and Treanor might well have refused and Vargas would have been unable to protect his solar panels from shading.

Would denying Vargas solar access protection have produced the efficient outcome in the case? The answer to that question depends on the relative values of the two competing uses of the airspace at issue. If protecting Vargas' panels from shading was worth $10,000 and allowing trees to grow in the airspace instead was worth $5,000, laws that resulted in no solar access easement would be efficient. Proponents of strong solar access laws tend to focus on this type of scenario, characterizing shading risk as an obstacle to valuable solar energy development and advocating laws that eliminate such risk.

On the other hand, suppose that allowing trees to grow in the airspace was worth $20,000 rather than just $5,000. Under this revised set of facts, aggressive laws like California's SSCA would have produced an inefficient result by giving Vargas a property interest in his neighbors' airspace even when Vargas' proposed use of it was not as valuable as theirs. As mentioned earlier, the false presumption that solar access protection is always the highest of all possible airspace uses underlies many aggressive solar access laws that automatically reallocate airspace rights among neighbors to prevent solar panel shading.

Incorporating a compensation requirement into a solar access statute largely spares policymakers from having to presume that solar access is or is not more valuable than other competing airspace uses. Instead, it compels landowners who seek solar access protection to reveal whether they truly value that protection more highly than the market values all other possible uses of the airspace at issue.

A simple numerical example helps to emphasize this point. Suppose that Vargas lived in Iowa rather than California and thus could compel Bissett and Treanor to grant him a solar access easement but had to *pay* them fair market value for it. If he were rational, Vargas would purchase the easement under such a rule only when it cost less than the value he would obtain from it. In other words, he would have purchased it when allowing trees in the space was worth $5,000 but not when allowing trees in the space was worth $20,000—the efficient outcome under both scenarios.

Table 5.1 below summarizes the advantages and disadvantages of each of the three general types of policy approach to solar access described in this chapter. "Policy type #1"—the predominant policy strategy for solar access throughout the world—is to have no special statutory or other laws that protect solar energy systems from shading. As shown in the table, this approach tends to promote the efficient use of neighbors' airspace whenever occupying that space with trees, buildings or other structures is of greater value than using it to deliver direct sunlight to a solar energy system. However, when keeping neighbors' airspace open to protect a solar energy device from shading would be the highest valued use of the space, this type of policy offers no sure way for would-be solar energy users to protect themselves against shading risks. If they are unable to convince neighbors to voluntarily sell them solar access easements, they have no other way

Table 5.1 Comparing the general efficiency of solar access laws

Types of solar access conflicts	Policy type #1: no statutory or common law solar access protection (approach in most jurisdictions)	Policy type #2: setbacks, permit-based protection, or solar rights statutes (used in some U.S. states[56])	Policy type #3: statutory options to purchase solar access easements at fair market value (applies in Iowa[57])
Conflicts in which solar access is *more* valuable than all competing airspace uses	Voluntary bargain required to reach efficient outcome	Efficient	Efficient
Conflicts in which solar access is *less* valuable than a competing airspace use	Efficient	Not efficient	Efficient

of preventing shading and can thus be deterred from using solar energy systems on their land.

"Policy type #2" in Table 5.1 encompasses those policies that essentially prohibit landowners from occupying their own airspace with trees and/or buildings when doing so would shade a neighbor's solar energy system. These aggressive pro-solar policies ensure cost-free solar access protection for all landowners who install solar energy devices on their land and are thus the most strongly favored by advocates of the solar energy industry. However, as described earlier in the chapter, such laws can produce inefficient outcomes whenever occupying the airspace at issue with trees, buildings, or other structures would be a higher valued use of the space.

Solar access policies modeled after Iowa's statute—"Policy type #3" in Table 5.1— require that parties who get solar access protection across neighboring airspace compensate neighbors for their consequent loss of airspace rights. This approach is not only more equitable to neighbors; it is also more likely to maximize the productivity of airspace. Landowners with solar energy systems must pay fair value for solar access systems rather than getting them for free under Iowa's rule, which incentivizes them to obtain solar access easements only when they value the space for that purpose more than the market values the space for other uses. Because of this unique feature, the Iowa approach is better able to promote the efficient use of airspace regardless of whether solar access protection is worth more or less than competing uses of the space.

The future of solar access laws

As distributed solar energy development continues to proliferate across the globe, conflicts over solar access will only increase. Although the ideas set forth in this chapter may help to govern these disputes, an even better solution would be to find ways to expand solar energy development while avoiding these conflicts altogether. Fortunately, some recent innovations in solar PV system design and trends in solar energy siting may ultimately help to do just that.

The most direct way to eliminate the solar access problem through technology would be to engineer solar panels that produced almost as much electricity in the shade as they produced in direct sunlight. There is no reason to believe that such a dramatic advancement will be made anytime soon, but researchers are gradually improving systems to make them less prone to shading. For example, a recent study found that the use of micro-inverters in solar panels significantly mitigated shade's impacts on solar panel productivity.[58] Hopefully, these technologies will continue to progress in the coming decades.

Another way to avoid solar access conflicts is to increase the proportion of solar energy development that takes the form of commercial- or utility-scale

projects rather than small rooftop solar arrays. These larger
to be sited on the flat rooftops of massive commercial buildir
far away from neighbors' trees or tall buildings, making then
trigger shading disputes. In the past decade, this category o
development has been rapidly expanding and, as would be
seemingly raised very few solar access disputes. Major ret......
Wal-Mart and Ikea have been adding solar energy generating capacity onto
their rooftops at an astounding pace.[59] Utility-scale solar energy facilities
have also been popping up in deserts at an unprecedented rate.[60] If this
trend continues, it could result in fewer and fewer solar access disputes per
installed kilowatt of solar energy generating capacity.

In summary, through innovations in the policies that govern solar access
conflicts, in solar energy systems themselves, and in project size, the future
for solar energy can be very bright in spite of the shade.

Notes

1 *See* Galen Barbose, et al., *Tracking the Sun V* at 2, Lawrence Berkeley
National Laboratory (November 2012), available at http://emp.lbl.gov/
publications/tracking-sun-v-historical-summary-installed-price-photo
voltaics-united-states-1998-2011 (noting that "installed prices" of solar PV
"exhibit significant economies of scale, with a median installed price of $7.7/W
for systems ≤2 kW completed in 2011, compared to $4.5/W for commercial
systems >1,000 kW").

2 *See* Ralf J. Muenster, *Shade Happens*, RenewableEnergyWorld.com (Feb. 3,
2009), available at www.renewableenergyworld.com/rea/news/article/2009/02/
shade-happens-54551 (last visited June 25, 2013).

3 *See, e.g.,* COLO. REV. STAT. §§ 38-32.5-100 et seq. For a list of these state
statutes, *see* Tawny L. Alvarez, *Don't Take My Sunshine Away: Right-to-Light
and Solar Energy in the Twenty-First Century*, 28 PACE L. REV. 535, 547
(footnote 90) (Spring 2008).

4 Portions of this chapter are adapted with changes from one of the author's
academic articles on the topic of solar access, Troy A. Rule, *Shadows on the
Cathedral: Solar Access Laws in a Different Light*, 2010 U. ILL. L. REV. 851
(2010).

5 Borimir Jordan & John Perlin, *Solar Energy Use and Litigation in Ancient
Times*, 1 SOLAR L. REP. 583, 592–93 (1979) (cited in Stephen Christopher Unger,
*Ancient Rights in Wrigleyville: An Argument for the Unobstructed View of a
National Pastime*, 38 IND. L. REV. 533, 542–43 (2005)).

6 *See* Sally J. McKee, *Solar Access Rights*, 23 URB. L. ANN. 437, 440 (1982). *See
also* BLACK'S LAW DICTIONARY (8th ed. 2004).

7 *See* McKee, *supra* note 6 at 400 (citing Aldred's Case, 77 Eng. Rep. (K.B. 1610),
as the first case to apply the doctrine).

8 *See* Prescription Act, 1832, 2 & 3 Will. 4, ch. 71, § 7 (1832) (cited in McKee,
supra note 6 at 440).

9 It should be noted that, as of early 2013, public discussion had begun in the
United Kingdom about the possibility of abolishing the doctrine of ancient
lights. These relatively new discussions are highlighted in more depth later in this
chapter. To download the United Kingdom Law Commission's full Consultation
Paper (No. 210) on the topic of "Rights to Light," visit the Law Commission

website at http://lawcommission.justice.gov.uk/consultations/rights-to-light.htm (last visited June 25, 2013).

10 *See* John William Gergacz, *Legal Aspects of Solar Energy: Statutory Approaches for Access to Sunlight*, 10 B.C. Envtl. Aff. L. Rev. 1, 6 (Fall 1982) (citing Gergacz, *Solar Energy Law: Easements of Access to Sunlight*, 10 N.M. L. Rev. 121, 141–44 (1980)) (identifying state courts in at least six U.S. states that had recognized the doctrine of ancient lights prior to 1939).

11 *See* McKee, *supra* note 6 at 441 (noting that "between 1835 and 1939 every state rejected" the doctrine of ancient lights "as unsuited to a developing country because it inhibited economically sound land use practices"). *See also Fontainebleau Hotel Corp. v. Forty-Five Twenty-Five, Inc.*, 114 So.2d 357 (1959) (holding that the "English doctrine of 'ancient lights' has been unanimously repudiated in this country").

12 For a more detailed description of the history of prescriptive sunlight access easement laws in Australia and New Zealand, *see generally Law Commission Consultation Paper No. 210: Rights to Light* at 31–33 (2013), available at http://lawcommission.justice.gov.uk/consultations/rights-to-light.htm (last visited June 25, 2013).

13 *See Lynch v. Hill*, 6 A.2d 614, 618 (1939) (cited in McKee, *supra* note 6 at 441) (declaring that the doctrine of ancient lights was "not suitable to the conditions of a new growing and populous country, which contains many large cities and towns, where buildings are often necessarily erected on small lots").

14 *See* Janice L. Dornbush, *Japanese Real Estate: Does the Government Help or Hinder Development?*, 5 Cornell Real Estate Rev. 1, 4 (2007).

15 *See* Bruce Yandle, *Grasping for the Heavens: 3-D Property Rights and the Global Commons*, 10 Duke Envtl. L. & Pol'y F. 13 (1999) (noting that "Japan underwent a construction boom in the 1960s" during which time many new high-rise buildings began to shade residential homes, causing an "upsurge in disputes concerning access to sunlight").

16 *See* F. G. Takagi, *Legal Protection of Solar Rights*, 10 Conn. L. Rev. 121, 136 (1977–78) (citing Miyazaki, *Nissho Kogaini Kansu Rukari Shobun Jiken* (Provisional disposition cases involving sunshine interference), 327 Hanrei Taimuzu 39 (1974)).

17 *See* Mohamed Boubekri, *Solar Access Legislation: A Historical Perspective of New York City and Tokyo*, Planning & Environmental Law, Vol. 57, No. 5, p. 3 (May 2005) (citing Mimpo, Law No. 89 of 1896 and Law No. 9 of 1998).

18 *See id.*

19 For a timeline describing in greater detail the history of solar energy technologies, including Edmund Becquerel's discovery of the photovoltaic effect in the 1800s and Bell Telephone Laboratories' development of the first silicon solar cell in 1954, *see generally The History of Solar*, U.S. Department of Energy: Energy Efficiency and Renewable Energy, available at http://www1.eere.energy.gov/solar/pdfs/solar_timeline.pdf (last visited June 26, 2013).

20 *See* Edna Sussman, *Reshaping Municipal and County Laws to Foster Green Building, Energy Efficiency, and Renewable Energy*, 165 N.Y.U. Envtl. L.J. 1, 30 (2008) (noting that, "[f]ollowing the oil embargo in the 1970s, there was a flurry of activity and legislation passed in various states addressing solar energy").

21 *See Prah v. Maretti*, 321 N.W.2d 182, 184 (1982).

22 *See id.* at 185.

23 Id. at 184.

24 Id. at 189–90.

25 Id. at 191.

26 *See* id. at 193.

27 *See* id. (quoting *United States v. Causby*, 328, U.S. 256, 264 (1946), for the principle that a "landowner owns at least as much of the space above the ground as he can occupy or use in connection with the land"); *see also* id. at 194 (asserting that the "right of a property owner to lawful enjoyment of his property should be vigorously protected").

28 *See* CAL. PUB. RES. CODE §§ 25980–25986 (2008).

29 Scott Anders, et al., *California's Solar Shade Control Act: A Review of the Statutes and Relevant Cases*, University of San Diego School of Law Energy Policy Initiatives Center (2010) at 1, available at www.sandiego.edu/documents/epic/100329_SSCA_Final_000.pdf (last visited June 27, 2013) (citing CAL. REV. & TAX CODE § 23601 (1976).

30 The California legislature amended to § 25983 of the statute in 2008 to remove its public nuisance provision, limiting liability to civil private nuisance claims. *See* CAL. PUB. RES. CODE § 25983. For a more detailed discussion of the "public nuisance" aspect of California's original Act, *see* Gergacz, *supra* note 10 at 24–25.

31 *See* Rule, *Sunlight on the Cathedral*, *supra* note 4 at 874 (citing the 1978 version of CAL. PUB. RES. CODE § 25983).

32 *See* Anders, *supra* note 29 at 10 (noting that "relatively few cases have examined the Solar Shade Control Act since its enactment" and citing just four cases citing the Act between 1978 and 2005).

33 *See* Douglas Fox, *Solar Energy Trumps Shade in California Prosecution of Tree Owner Across a Backyard Fence: When is the Environmentalism Greener on the Other Side?*, CHRISTIAN SCIENCE MONITOR, March 18, 2008 (available at www.csmonitor.com/2008/0318/p20s01-lihc.html (last visited June 28, 2013).

34 *See* Felicity Barringer, *Trees Block Solar Panels, and a Feud Ends in Court*, NEW YORK TIMES (Apr. 7, 2008), available at www.nytimes.com/2008/04/07/science/earth/07redwood.html?pagewanted=all&_r=0 (last visited June 28, 2013).

35 Fox, *supra* note 33.

36 Barringer, *supra* note 34.

37 Id.

38 *See* Anders, *supra* note 29 at 10.

39 *See* Barringer, *supra* note 34.

40 *See* David Dockter, *The "Trees vs. Solar" Issue Put to Rest in the Capital*, California Urban Forests Council, available at http://energycenter.org (last visited June 28, 2013) (observing that the "national news impressions and fallout from onlookers" to the Bissett-Vargas dispute "ranged from bewilderment, to humor to serious polarization").

41 *See, e.g.*, Marla Dickerson, *Hey, Your Shade Trees Are Blocking My Solar Panels*, LOS ANGELES TIMES (Nov. 15, 2008) (describing a similar shading dispute between neighboring landowners in Culver City, California).

42 2008 CAL. STAT. ch. 176.

43 *See* Anders, *supra* note 29 at 6.

44 *See* id. at 10.

45 *See* MASS. GEN. LAWS ANN. ch. 40A, § 9B (Lexis Nexis 2006).

46 IOWA CODE ANN. § 564A (West 1992).

47 *See* id. at § 564.A.4(1).

48 Iowa's statute specifically allows for voluntarily-negotiated solar access easements. *See* IOWA CODE ANN. § 564.A.7.

49 *See* IOWA CODE ANN. § 564.A.4.2.

50 *See* IOWA CODE ANN. § 564.A.4.3.

51 *See* IOWA CODE ANN. § 564.A.5.1.

52 Id.

53 Iowa Code Ann. § 564.A.5.3.

54 *See* Iowa Code Ann. § 564.A.5.4.

55 *See* Iowa Code Ann. § 564.A.6.

56 *See, e.g.*, N.M. Stat. § 47-3-4 (1978); Wyo.Stat. Ann. §§ 34-22-101 to 104 (2009); Mass. Ann. Laws ch. 140A, § 9B (Lexis Nexis 2006).

57 *See* Iowa Code Ann. § 564A.

58 *See, e.g.*, Chris Deline, et al., *Photovoltaic Shading Testbed for Module-Level Power Electronics* at 1, National Renewable Energy Laboratory (May 2012) (noting that using a microinverter on an 8-kW solar PV array increased its energy productivity by 12.3 percent under "heavy shading" conditions).

59 *See generally Solar Means Business: Top Commercial Solar Customers in the U.S.*, Solar Energy Industries Association (Nov. 2012), available at www.seia. org/research-resources/solar-means-business-top-commercial-solar-customers-us (last visited June 30, 2013) (describing rapid growth in solar energy installations by large retail companies).

60 *See generally* Philip Wolfe, *The Rise of Utility-Scale Solar*, RenewableEnergyWorld. com (April 15, 2013), available at www.renewableenergyworld.com/rea/news/ article/2013/04/the-rise-and-rise-of-utility-scale-solar (last visited July 2, 2013) (noting that "utility-scale power plants are now the fastest-growing application for the photovoltaics (PV) industry").

6 Renewable energy vs. cultural preservation

As illustrated in the previous two chapters, many opponents of renewable energy projects are generally supportive of wind and solar energy but happen to believe that some other ideal—saving trees, protecting wildlife, etc.—should take precedence over a particular project. Indeed, many conflicts over renewable energy development are ultimately clashes between competing interests that are each noble and legitimate in their own right. This chapter focuses on yet another important social policy goal that can complicate renewable energy siting: cultural and historic preservation.

The basic premise underlying most historic preservation efforts is that society should protect and preserve cultural and historic resources for the benefit of future generations. Historically-significant sites and artifacts can be powerful educational tools, vivid reminders of a society's past failures and successes, and precious contributors to a nation's social fabric. Destroying or altering cultural and historic resources to serve short-term interests imposes costs on generations yet unborn who might have otherwise benefited from witnessing those resources in their undisturbed condition. In that sense, preserving historic resources is a gift to the planet's future inhabitants, enabling them to extract value from significant sites and cultural items for centuries to come.

Of course, new renewable energy projects can clearly benefit future generations as well.[1] New wind farms and solar arrays reduce the quantity of fossil fuels that utilities must burn to meet customers' energy needs. By displacing fossil fuel combustion, renewable energy can slow the pace at which such combustion impairs global ecosystems and exhausts the planet's finite supply of nonrenewable energy resources. Renewable energy projects can also improve a nation's long-term economic security: once installed, a solar energy system or wind turbine can generate clean, usable electric power for decades with relatively minimal maintenance. For these and other reasons, new wind farms and solar arrays can likewise confer valuable benefits to those who will occupy this planet decades or even centuries from now.

Unfortunately, the competing ideals of historic preservation and sustainable energy can come into conflict whenever a newly proposed renewable energy project threatens to harm a precious cultural resource.

Wind and solar energy development can intrude upon tribal burial grounds, disturb views of religiously significant mountains, or blemish the famed rooftops of vintage buildings. Even though such development may help keep the earth clean and livable for future generations, it can also permanently mar or destroy priceless cultural assets.

As stewards of the earth—its environmental resources *and* its cultural treasures—humankind is increasingly struggling to balance the goals of sustainable energy and cultural preservation. Should government officials deny a wind farm proposal on public land just because a few arrowheads are found nearby? Should laws prohibit rooftop solar panel installations on a privately owned building solely because the solar array might detract from the authentic look and feel of a registered historic district? What principles should guide policymakers throughout the world as they seek to balance green energy and historic preservation goals in their own jurisdictions? And what groups should be empowered to make these difficult balancing decisions?

This chapter explores the intrinsic tension between sustainable energy and cultural preservation and offers some general thoughts on how governments might better address this tension as renewable energy projects encroach more and more into historically significant areas.

Obstacles to renewable energy development on tribal lands

One set of policy challenges at the intersection of renewable energy and cultural preservation involves situations when groups seek to develop wind or solar energy projects on tribal lands.

Some of the world's most promising wind and solar resources are situated on lands controlled or occupied by indigenous groups. Many such lands are in relatively isolated regions, far removed from major population centers. In the Americas and in Australia, colonists and homesteaders largely avoided these areas for centuries, concluding that they were too rugged or out-of-the-way to be valuable for commerce or agriculture. For a variety of reasons, governments ultimately set aside some of these seemingly less-desirable areas for occupation by indigenous peoples[2]—groups of pre-colonial inhabitants that are commonly referred to as Indians in the United States and some other Western countries. Often, this involuntary displacement process removed Indians from many of their ancestral lands, relocating them to "reservation" lands held by governments in trust for tribal use.[3]

Over the past century, huge stores of fossil fuels have been discovered on some Indian trust lands and mining or extraction operations have followed. For example, major coal mines have been active on Navajo nation lands in Arizona in the United States for nearly a half century, supplying millions of tons of coal per year to a nearby electric generating station.[4] Although mining and extraction projects can provide employment opportunities and income to tribes, they can also alter the landscapes of reservation lands and bring environmental and health hazards with them. Understandably, these bittersweet arrangements have fostered uneasy love–hate relationships between the energy industry and some tribes.

As renewable energy technologies have advanced in the past decade, developers and policymakers have once again begun looking more earnestly to Indian trust lands for energy resources. These reservation lands, once deemed too rugged and remote for profitable conventional development, are now garnering attention as valuable potential wind and solar energy sites. According to one estimate, wind energy resources on Indian trust lands in the continental United States alone are capable of producing as much as 535 billion kWh of electric power per year—an amount equal to 14 percent of all U.S. electricity generation.[5] Solar energy resources on U.S. Indian trust lands could produce up to 17,600 billion kWh of electricity per year, or roughly 4.5 times the total of all of that country's electric generation in 2004.[6]

Wind and solar energy development on tribal lands can potentially do much to improve the lives of indigenous groups. It can provide additional revenue to tribes, whether in the form of lease payments from outside developers or direct electricity sales revenues for tribes that develop projects on their own. New employment opportunities also often accompany these projects, and such jobs tend to pay decent salaries. And renewable energy systems can potentially electrify homes and communities that are currently in the dark on Indian reservations in far-flung areas that have limited access to the power grid.[7]

Renewable energy development is also arguably more compatible with the values and belief structures of many indigenous groups than other common forms of tribal revenue-raising such as casinos and coal mines. Although there is great diversity among indigenous beliefs,[8] some common threads relating to the natural environment seem to run through most such cultures. In the words of one scholar, "most native peoples do not think of nature in exclusively economic terms, as a commodity to be exploited at will."[9] Instead, they place exceptionally high value on sustainable living and harmony with nature.[10] For obvious reasons, renewable energy strategies seem to better respect such harmony than strategies that rely on the extraction and burning of fossil fuels. The sustainable and more ecologically-friendly characteristics of renewable energy development can thus make these projects a particularly appealing enterprise for some tribes.[11]

Even wind and solar resources themselves are revered in some indigenous cultures, which can sometimes make these groups particularly open to renewable energy development. For example, the sun has religious significance among several Indian tribes in the United States. According to a researcher of this topic:

> it is a common characteristic among American Indians to revere the sun and to value its energy-creating capacities—whether it is for the

production of crops, the signaling of weather change, how long to stay at sea for purposes of subsistence fishing, the movement of game, or simply as a means of entertainment. This veneration often runs deep ... As a result, tribes are finding themselves at the forefront of the renewable energy trend, and, as ancestral stewards of the earth, are embracing alternative energy resources on their land.[12]

Unfortunately, as more and more American Indian tribes have explored the possibilities of renewable energy development on their lands, the challenges associated with it have grown increasingly apparent. Within tribes, disputes can arise over the mere prospect of renewable energy development on tribal lands. Despite the economic opportunities such development could bring, some tribe members may staunchly oppose it on the simple basis that tribal lands are too sacred to be sites of industrial-scale development.[13]

Even when the majority of a tribe's members are supportive of a proposed renewable energy project on tribal land, other complications can muddle and slow the development process. For example, it is not always clear to renewable developers with whom they should negotiate the terms of lease agreements on Indian trust lands. In the United States, title to these reservation lands is typically vested in the federal government, necessitating the involvement of government agency officials. Still, how much weight should tribal input receive in these discussions, and who within a tribe holds authority to provide it? As one commentator has noted,[14]

> [W]ithin reservations, the decentralization and dispersal of decision making authority may cause developers fits. In one case involving proposed wind energy on an Indian reservation, one project proponent obtained rights from the local Indian community, another obtained commitments from the President of the Indian nation for the same land, and yet another was seeking the same development rights from influential members of the tribal legislature.

Even in the absence of inner-tribal squabbles or questions of authority, renewable energy development tends to be more complicated *on* the reservation than *off* of reservation lands. This increased complexity stems in part from the fact that private developers of off-reservation projects can often more easily take advantage of subsidy programs that do not apply as readily to indigenous groups. For instance, as of 2013, some federal tax credit programs aimed at incentivizing commercial wind energy development in the United States were arguably more difficult for tribal groups to participate in because tribes are not federal taxpayers and thus cannot as easily avail themselves of such credits.[15] For tribes located in remote areas, costly expansions of existing high-voltage transmission lines are also needed to make any major renewable energy project feasible because the existing transmission infrastructure was not designed with on-reservation energy generation in mind.[16]

To date, policies in the United States aimed at overcoming barriers to renewable energy on Indian trust lands have largely proven ineffective. The United States Congress enacted the Indian Tribal Development and Self-Determination Act (ITEDSA) in 2005, seeking to ease energy development on reservation lands.[17] In connection with this legislation, the federal government formed new agencies specifically designed to assist tribes with energy development.[18] Unfortunately, ITEDSA's provisions failed to adequately address several remaining obstacles to renewable energy projects on tribal lands, including the challenges described above. For instance, as of late 2013, not a single tribe had taken advantage of certain ITEDSA provisions allowing tribes to obtain greater autonomy in tribal energy development through special Tribal Energy Resource Agreements (TERAs).[19] A primary reason for this lack of tribe participation is that few tribes have adequate financial resources to fund their involvement in the TERA application process.

Members of Congress have advocated for new legislation in recent years that would fill some of the holes left by ITEDSA, but as of late 2013 their efforts on a new bill had proven unsuccessful.[20] Frustrated by this lack of progress, scholars and stakeholders in the United States have been increasingly advocating policy changes capable of creating a more level playing field and greater opportunities for Indian tribes interested in developing renewable energy projects on reservation lands.[21]

Despite all of the aforementioned challenges, a small handful of indigenous groups throughout the world have been finding ways to develop wind and solar energy facilities on tribal lands, and are having some limited success. For example, the Moapa tribe in southern Nevada in the United States has received approvals and begun developing a 350 MW solar energy project on its territory.[22] The project has been touted as the first ever utility-scale solar energy project on tribal lands in the United States.[23] Developers in Australia also signed a lease agreement in 2011 for wind energy development on Aboriginal lands in the state of Queensland.[24] If built, the project would be the first wind farm on Aboriginal land in that country.[25] Hopefully, these pioneering projects will prove successful and will open the door for much more responsible, mutually beneficial renewable energy development on indigenous lands over the next century.

Tribal opposition to renewable energy projects outside the reservation

Although many Indian tribes in the United States are interested in the potential economic rewards of pursuing renewable energy development within their Indian Trust lands, tribes have increasingly taken an opposite stance with respect to *off*-reservation projects. Tribes can be fierce opponents of wind and solar energy projects from which they stand to gain no financial benefit, particularly when such projects are on ancestral lands of cultural significance to them. On multiple occasions, challenges from tribal groups have been a major obstacle to renewable energy development in these areas.

Indigenous groups' objections to renewable energy projects are commonly rooted in concerns that the project could harm sacred sites, artifacts, or other cultural resources. In the United States and Australia, indigenous peoples control far less land today than they did prior to colonization and relocation. Often, the reservations or territories where tribes reside constitute much smaller geographic areas than those that they occupied and used for centuries before colonists arrived. Much of the land on which these tribes once hunted, engaged in religious worship, and even buried their deceased is no longer under their control. When a wind or solar energy project proposed outside of Indian trust lands threatens to damage historic elements of a tribe's ancestral areas or to disturb sacred sites, it is easy to understand why the tribe might oppose such a plan.

Vast areas of remote land in the western United States that were occupied by Indian tribes prior to colonization are increasingly being targeted for renewable energy development. More than 650 million acres of U.S. land—roughly 30 percent of the entire country's land mass—are designated as federal public lands. The federal government assumes primary responsibility for managing and maintaining this land for the benefit of the citizenry. In an attempt to aggressively promote renewable energy development on public lands, the Obama administration adopted policies in 2011 designed to expedite the permitting and approval process for certain selected projects.[26] As of August 2013, dozens of wind and solar energy projects on lands managed by the U.S. Bureau of Land Management (BLM) had received approval or were under consideration.[27] Some of these fast-tracked projects are near Indian reservations or on lands that are culturally significant to Indian groups.

Wind and solar energy development may be more environmentally friendly on average than coal- or oil-based energy, but it can still cause severe damage to cultural resources. The BLM itself concedes that utility-scale solar energy projects can adversely affect "cultural resources, including historic properties ... regardless of the technology employed."[28] Among other things, the agency has noted that noise from solar energy projects can potentially interfere with "spiritual, religious, and medical practices and ceremonies [that] are ongoing within the desert southwest."[29] The BLM has similarly acknowledged the potential for wind farms to harm sacred Indian sites or other cultural assets. Wind energy projects inevitably involve new roads, transmission lines, and related disturbances to the natural terrain. As the BLM has stated, these projects can have disruptive "[v]isual impacts on ... sacred landscapes, historic trails, and viewsheds from other types of historic properties."[30]

Because of the potential for development on public lands to adversely impact culturally significant tribal resources, laws in the United States seek to minimize such harms by encouraging open discussion between developers and tribes. Federal laws in the United States require BLM officials contemplating development projects on public lands to "consult" with any American Indian tribes that "attach [...] religious or cultural significance" to the lands involved.[31] Unfortunately, disagreement regarding what these laws require can sometimes leave tribes unsatisfied with developers' consultation efforts, and costly lawsuits can follow.

Clashes over tribe consultation requirements in the southwest United States

Multiple Indian tribes in the United States have opposed proposals for renewable energy projects on public lands in efforts to protect their cultural resources. One example of such opposition involves the Quechan Tribe of the Fort Yuma Reservation (Quechan Tribe) in southern California. Over the course of only a few years, the Quechan Tribe challenged two different renewable energy projects based on claims that BLM officials failed to adequately consult with them as required under Section 106 of the National Historic Preservation Act (NHPA) and its accompanying regulations.[32]

The Quechan Tribe's first major clash with federal officials over the NHPA's consultation requirements arose in 2010, when the BLM approved a large solar energy project on roughly 6,500 acres of mostly public lands in the California Desert Conservation Area in Imperial County, California.[33] The proposed 700+ MW project involved land that was home to 459 identified cultural resources of various kinds.[34] At least 300 of these resources were "locations of prehistoric use or settlement," including archeological and burial sites. Many resources falling within this category were culturally significant to the Quechan Tribe, whose ancestral lands included portions of the proposed project area.[35]

Cognizant of its obligations under the NHPA, the BLM held town hall meetings and consultations with individual tribal members and provided written updates to the Quechan Tribe about the solar energy project.[36] However, the project was moving forward at an unusually rapid pace in order to meet a critical deadline necessary to qualify for certain federal stimulus funds. Resenting what it felt was a cursory approach to the consultation process in this rush to meet these deadlines, the tribe filed suit against the BLM and sought to enjoin the project's construction on the ground that the BLM's consultation efforts were inadequate.

In December of 2010, a federal district court in California issued an opinion that largely agreed with the Quechan Tribe's criticisms of the consultation process for the project. The court held that the BLM's interactions with the tribe did not "amount to the type of 'government-to-government' consultation required by federal law."[37] In explaining its holding, the court gave some guidance regarding what federal officials must do to comply with NHPA's consultation requirement in connection with renewable energy projects on public lands. Among other things, federal officials must "begin early" in consulting with tribes about large energy projects on public lands to ensure that tribes have "adequate information and time" to investigate the project and its potential impacts on cultural resources.[38] The court's decision was a significant victory for the tribe, offering hope that siting consultations with federal officials would be more collaborative on future projects.

Shortly after prevailing in court against this *solar* energy project, the Quechan Tribe filed suit seeking to enjoin the development of a *wind* energy project within its ancestral lands on similar grounds. Pattern Energy Group's Ocotillo Express Wind project was initially designed to involve 155 commercial-scale wind turbines covering roughly 13,000 acres of public land in Imperial County, California.[39] The BLM and Pattern Energy Group were surely aware of the tribe's lawsuit against the solar project, which may have motivated them to be extra cautious and thorough in the consultation process. They ultimately trimmed back the project to just 112 turbines covering about 10,000 acres in response to input and information gathered through consultations with tribes and environmental review.[40] Among other things, these changes to the project plans were aimed at avoiding the siting of turbines where they might disrupt certain viewsheds that were significant to the several tribes that were consulted about the proposal.[41]

Dissatisfied with Pattern Energy Group's revisions to the project plans, the Quechan Tribe filed a claim in federal court in May of 2012 seeking to enjoin the project. The primary legal argument in the tribe's claim mirrored that of its prior lawsuit: that the BLM had failed to adequately consult the tribe. According to the complaint, the project area contained "hundreds of archeological sites (containing tens of thousands of individual artifacts) eligible and potentially eligible for inclusion in the National Register of Historic Places" and held "tremendous spiritual essence for the Quechan Tribe."[42]

Unfortunately for the Quechan Tribe, the court was less sympathetic to its arguments this time around. The court dismissed the tribe's claim, finding that the BLM's administrative record evidenced that the agency had made "many attempts, starting regularly in 2010 … to engage the Tribe in Section 106 government-to-government consultation."[43] According to the court, the tribe "did not accept these repeated requests until December 2011, towards the end of the approval process," at which point its fair opportunity to challenge the project had passed.[44]

After the court issued its opinion, construction on the Ocotillo Express Wind Project proceeded. However, as of late 2013, the legal fight over the Ocotillo Express Wind Project was not entirely over. The Quechan Tribe appealed its case and, in April of 2013, the California Native American Heritage Commission declared the entire project area to be a sacred Native American site.[45] The tribe has also challenged even more renewable energy projects in the area based on inadequate consultations.[46]

The Quechan Tribe is by no means the only Indian tribe that has questioned the adequacy of BLM consultation efforts in recent years. Another example of such a dispute involves the Colorado River Indian tribes in western Arizona and southern California, who made similar arguments in 2012 against a solar energy project proposed on public land in California's famed Mojave Desert.[47] Even though the proposed Genesis project was off of the tribes' reservation area, tribal leaders argued that ancestral remains found on the land were part of a "living spiritual world" and that to disturb them would "disrupt [...] the peace of [their] ancestors."[48] The tribes also sent a letter to the California Energy Commission in January of 2013 complaining that the BLM and other government agencies had "failed to take seriously concerns raised by the tribes."[49] The letter alleged that agency officials were unacceptably placing renewable energy goals above that of preserving cultural resources, citing the agency's own admissions that "[c]ultural resources are non-renewable and, once damaged or destroyed, are not recoverable."[50] Unfortunately for the tribes, project construction continued in spite of their pleas.

These ongoing disputes highlight the conflict that often arises when renewable projects are proposed on public lands that have cultural significance to Indian tribes in the United States. As mentioned above, this tension is partly attributable to ambiguity regarding what constitutes an adequate NHPA Section 106 "consultation" between public officials and interested tribal groups in this context.

Some tribal rights advocates interpret the NHPA's consultation requirement as effectively requiring federal officials to accede to tribal demands regarding development on public lands. Indeed, a guide published by the Indigenous Peoples Subcommittee of the EPA's National Environmental Justice Advisory Council distinguishes "consultation" requirements from less onerous "public participation" requirements and suggests that consultation necessitates at least some deference to tribal wishes. According to the guide:

> Conventional *public participation* initiatives ... [do] not require the federal government to change its decision based on localized, public input ... On the other hand, consultation between the federal and tribal governments should be a collaborative process between government peers that seeks to reach a consensus on how to proceed ... [I]n some instances, specific requirements demand the federal government give special deference to tribal preference.[51]

The guide later elaborates that consultation "generally means much more than sending letters, notices, and copies of documents to tribes and requesting comment."[52] This language arguably implies a viewpoint that, to at least some degree, federal agencies must acquiesce to tribes' restrictions on development on public lands where tribal cultural resources are at stake, even if such conditions hinder a valuable renewable energy project.

Not surprisingly, the BLM and other government agencies seeking to promote renewable energy on public lands tend to take a somewhat different view of the consultation requirement. From their perspective, their duty to consult with tribes is not an obligation to terminate or amend proposed development plans on off-reservation lands based on tribal input. Nothing in the NHPA or its implementing regulations is "meant to suggest federal agencies must acquiesce to every tribal request."[53] Thus, as a BLM spokeswoman speaking about the Ocotillo Express Wind Project explained in connection with that project, NHPA's consultation requirement "doesn't necessarily require agreement all of the time."[54]

Somewhere between these rival interpretations of the NHPA's consultation requirement there exists an optimal middle ground—a level of respect for tribal concerns that efficiently and fairly balances the competing interests of renewable energy development and cultural preservation. Unfortunately, because inquiries into what constitutes an adequate consultation under NHPA can be very fact-specific, the struggle to find this balance is likely to continue to play out in courts well into the future. In the face of this uncertainty, developers of energy projects on public U.S. lands would be wise to reach out early and often in the planning stages to all indigenous groups whose cultural resources may be affected by their projects. The shrewdest of developers recognize the value of meticulously documenting their efforts to engage with and genuinely address the concerns of affected tribes throughout the planning and development process.

Wind farms vs. indigenous groups in southern Mexico

Although clashes between renewable energy development and Indian tribes in the United States tend to garner more media attention, similar conflicts are increasingly arising outside of the United States as well. On occasion, Indian tribes in Canada have taken aim at proposed wind energy projects based on their potential impacts on cultural resources.[55] As the following paragraphs describe, struggles between wind energy developers and indigenous groups are also arising south of the U.S. border. Although legal landscapes vary from nation to nation, the basic challenges and questions in these conflicts are largely the same: to what extent should laws restrict renewable energy development to protect the cultural heritage and interests of indigenous peoples?

Tensions between renewable energy and indigenous groups have been boiling for several years along the Isthmus of Tehuantepec in the southern Mexican state of Oaxaca—a region renowned for its prime wind energy resources. The area has long been called "La Ventosa," which is Spanish for "The Windy."[56] This narrow strip of land separating the Pacific Ocean and the Gulf of Mexico has been described as a global "wind hotspot"[57] and as "one of the windiest places on Earth."[58] As of late 2012, the Isthmus of Tehuantepec was home to at least 14 wind energy projects and more than

400 wind turbines, with several other projects in various planning stages by developers owned by such companies as Mitsubishi and Macquarie.[59] Wind energy development in the region has been proceeding at a breakneck pace in recent years, including a 140 percent year-over-year increase in generating capacity in 2012 alone.[60]

The Isthmus of Tehuantepec is a relatively rural area of Mexico, occupied in large part by impoverished indigenous groups and peasants who rely on fishing and low-technology agriculture for subsistence. Many of these citizens hold land collectively in *ejidos*, which generally means that some sort of group consent is required to convey any real property interests.[61] Ignoring this communal ownership structure, some developers have entered into wind energy lease agreements with occupiers of these areas that treat these residents as independent property owners. The rental amounts under many of these agreements are also extremely unfavorable to lessors—as little as 100 Mexican pesos[62] (less than eight U.S. dollars[63]) per hectare per year. Advocates for these groups have denounced such rental practices, arguing that they "violate collective property rights, agrarian laws and the traditional laws of indigenous peoples."[64]

Indigenous groups opposing wind energy development in this region of southern Mexico are concerned about a different type of cultural preservation than the sort that generally troubles tribes north of the U.S.–Mexican border. The concerns of the indigenous peoples of Oaxaca are not centered on possible disturbances to sacred artifacts and burial sites. Instead, these people are seeking to protect themselves against direct threats to their centuries-old, traditional lifestyle. Wind turbines and supporting facilities encroaching into the remote areas that these tribes have long relied on for their simple subsistence can put the very livelihood of these vulnerable groups at risk.

Disharmony between wind energy developers and indigenous locals escalated in 2012 in connection with the San Dionisio Del Mar wind energy project—a project planned in an area near Oaxaca's western coastline that aimed to be the largest wind farm in the country.[65] Many residents living in the project area are members of the "Ikojts" or "Ikoots" group, a tribe that has relied primarily on fishing and small-scale agriculture to sustain its way of life in the area for centuries.[66] As construction began on the project, area residents grew increasingly convinced that the project developer had ignored locals' longstanding *ejido* property ownership structure and coaxed them into signing unfair, one-sided contracts. Residents were concerned that the wind farm would "desecrate Ikoots sacred territory", including the picturesque Isla de San Dionisio and Barra de Santa Teresa.[67] And some locals were also furious over the developer's alleged practice of committing to pay large sums of money to influential local leaders to secure their support for the project and then offering far less compensation to all other residents.[68] Some feared that royalties payable to common residents under those contracts would not adequately compensate them for the project's impacts on their livelihood. In the words of one local resident:

This is the life of the poor: we fish so we can eat and have something to sell, to have a bit of money. They say that now that the wind project is here, they'll give us money for our land and sea, but the money won't last forever. We don't agree with this. How are we going to live?[69]

Unable to peaceably resolve their grievances and increasingly desperate to preserve their way of life, the impoverished indigenous peoples of San Dionisio del Mar grew ever more hostile as the development project progressed. According to one account, the group stormed the town's municipal palace in January of 2012, ousting the municipal president and replacing its regime with a general assembly that was bent on preventing the wind project from going forward.[70] The assembly then established a "permanent watch" to keep the developer's construction contractors out of the town. Acts of violence and intimidation ensued, ranging from blockades to stoning to attacks on the street. According to one local newspaper, at least 14 violent events relating to the wind farm project occurred in the last four months of 2012 alone.[71]

Determined to stop the wind farm project, the Ikoots and other locals also took its developer to court. Recognizing that the developer had not adequately secured rights to build the project under local law, a Mexican federal court granted the group's request for a temporary injunction. Among other things, the court found that the developer's actions in connection with the project were "in violation of the land rights of the community."[72] Unfortunately, the injunction proved a short-lived victory for the Ikoots people. The developer somehow managed to resume project construction work within a few months amid death threats aimed at vocal opponents of the project, drawing criticism from some justice advocacy groups.[73]

The unrest surrounding the San Dionisio Del Mar Wind Project is emblematic of an ongoing struggle between wind energy developers and indigenous groups throughout southern Mexico. Opposition to the region's Piedra Larga I wind farm in the city of Union Hidalgo was tied to the fatal shooting of a wind energy contractor near that community in 2011.[74] When developers sought to commence construction on Piedra Larga II, a local "Resistance Committee" comprised mostly of the community's native Zapotec people soon formed. The group traveled to Mexico City in mid-2013 to demand that their contracts with the developer be nullified and that construction immediately cease.

In a report before Mexico's Agrarian Unitary Tribunal, the opposition group denounced what they considered to be "lies, criminalization of protests, pollution of lands and waters, [and] deaths of birds and land animals" associated with the wind project's first phase.[75] Based on their experience with Phase I of the Piedra Larga project, they were determined to prevent construction of the second phase. In the words of one group member, "wind-energy is not clean energy for us; instead, it is an energy of death."[76]

As illustrated by these sorts of ongoing protests, tribal opposition to wind energy remains strong in some parts of the blustery Isthmus of Tehuantepec region. If nothing else, hopefully the indigenous peoples of southern Mexico are growing more aware of the potential consequences of entering into agreements with wind energy developers. Such awareness and the additional caution it could bring to their negotiations might do something to help to limit future upheavals in this windy corner of the world. Meanwhile, one can only hope for less violence and more equitable treatment of local residents as development of the uniquely abundant wind energy resources in this area continues.

New Zealand's Maori people and claims of ownership in the wind

One other interesting conflict involving renewable energy and tribal cultural resources centers on a tribe's recent assertion that it held a property interest in the wind flying over an entire country. This unusual dispute highlights how legal uncertainty about the property rights associated with wind energy development continues to unnecessarily complicate development efforts in this important industry.

What if the "cultural resource" at issue in a wind energy development conflict were the wind itself? Not long ago, a Maori tribe in New Zealand made such a claim. When New Zealand began preparing to partially privatize its electric utilities, the Maori gained some traction in arguing that, under the terms of an old treaty, the tribe held some limited rights in the nation's water resources.[77] On similar reasoning, spokesman David Rankin of the Ngapuhi tribe created a small media stir in 2012 by declaring that his tribe also had a special property interest in New Zealand's wind.

Rankin based the tribe's controversial argument on the idea that the wind had long been worshipped under the Maori traditional belief system. Given the Maori people's long-held reverence for wind, Rankin suggested it would violate their rights if New Zealand allowed its exploitation for commercial gain without permission. In Rankin's words, "Traditionally, the wind was regarded as a deity in Maori society, and Maori do not consider the Crown to have the right to use it without Maori consent."[78]

Rankin's reasoning seemed to be that the Maori held an ownership interest in the wind in the same sense that they allegedly held rights in water resources. "They're all resources," Mr. Rankin told a reporter. "The wind is a resource. What turns those turbines on those wind farms?"[79] A potential for financial gain seemed to at least partly motivate the group's newfound attempts to stake property claims in the wind.[80] Interestingly, Rankin said that Maori claims in airspace could be next.[81]

Still, although the Maori's assertions of property rights in wind seemed somewhat opportunistic, the group's religious reverence for wind is real. When Genesis Energy proposed its Castle Hill Wind Farm, a massive wind energy project north of Masterson, New Zealand, Maori groups in the area prepared a report outlining its potential impacts on cultural resources. Included in the report was a discussion of wind in the context of the groups' long-held beliefs.[82] According to the report:

> The wind has been used for recreation. The game with respect to wind is *manuaute*, kite flying. While this is something that can happen all year round, Maori kite flying has been the focus of Matariki. The Maori New Year was an opportunity to remember the past and celebrate the future by enjoying games like kite flying. When Whatonga arrived in this new land he studied the skies and the winds and as he played with the *manuaute* he realised the power of the wind ... The wind is a source of power that has allowed *tipuna* to establish our people in this land.[83]

The report further declared that any "[p]ermanent change of the landscape as a consequence of constructing wind farms is for Maori a loss of *mana*.[84] In the Maori language, *mana* means "status, esteem, prestige or authority."[85]

In spite of these arguments, it appears that the New Zealand government will not recognize any proprietary interest in the wind in favor of the Maori people. Consistent with this position, the Castle Hill Wind Farm received its resource consents from the New Zealand government in mid-2013.[86] The Maori people will undoubtedly continue to worship the wind, but their attempts to leverage that reverence into an economic interest ultimately proved unsuccessful.

Opportunities for collaboration

Despite the various clashes that have arisen between renewable energy developers and indigenous peoples in recent years, there is reason to believe that mutually beneficial partnerships are possible between these two groups. Given the vast quantity of developable renewable energy sites on tribal lands and tribal values centered on sustainability, there are plenty of reasons to improve their partnership. Renewable energy development on or near reservations has the potential to improve the lives of millions of indigenous individuals and to make a significant contribution toward the greening of the global energy economy. As coal-fired power plants on or near reservation lands are decommissioned in the decades to come, transmission capacity capable of delivering renewable energy from Indian trust lands to major population centers could become more available to tribes in some areas.

However, renewable energy growth on and near reservation lands will only become a reality if policymakers place a greater focus on encouraging cooperation between renewable energy interests and tribal interests. What, if anything, can policymakers do to improve the current state of relations between tribes, developers, and governments?

One possible means of improving this relationship would be through a policy that somehow gave tribes some "skin in the game" in connection with renewable energy projects on off-reservation public lands. Presently, projects on off-reservation ancestral lands are a "lose–lose" for tribes: these projects threaten a tribe's cultural resources yet offer no prospect of any economic benefit. A conceivable way of addressing this problem would be a program that financially rewards tribes for identifying off-reservation areas within their ancestral lands on which they would not object to renewable energy development.

For example, suppose that the United States enacted legislation under which a tribe could identify "pre-approval zones" on off-reservation, public ancestral lands within which the tribe would agree not to oppose renewable energy development under certain specified conditions. In exchange, the United States government would agree to distribute to the tribe a modest percentage of all lease revenue the government receives from renewable energy leasing within these "pre-approval zones." If necessary, the government could distribute these payments to tribes in the form of aid. The aid could even be designed to help fund installations of small solar energy systems and small- to medium-sized wind turbines capable of supplying power to tribe members that currently lacked grid access.

Under a program like the one just described, tribes could potentially receive real economic benefits when renewable energy projects were developed on off-reservation public lands that were part of their ancestral lands. This prospect of an economic benefit for tribes might promote more amicable relations and greater collaboration between the BLM, developers, and tribes, enabling more of these projects.

Likewise, there are ways that policymakers could improve tribes' ability to develop wind and solar energy projects *on* reservations. Even if the United States government proves slow to adapt policies that promote on-reservation renewable energy projects, state governments could do much in this regard. For example, state legislatures in the United States could add "tribal renewable energy" carve-outs to their renewable portfolio standard (RPS) legislation, requiring that utilities obtain some proportion of their renewable power from on-reservation renewable energy projects to satisfy the state's RPS goals. Under such provisions, the fact that a given unit of renewable energy was generated on Indian trust lands within a state would be an additional "attribute" capable of attaching to renewable energy credits. By effectively mandating that utilities obtain some of their renewable power from projects on reservations, this sort of policy would create greater demand for on-reservation wind and solar energy.[87]

In summary, much could be done to bridge differences between renewable energy interests and tribes and thereby remove unnecessary obstacles that are slowing the sustainable energy movement. Admittedly, political support for these sorts of policies presently seems weak in many countries, so it is doubtful that such policies will be adopted anytime soon. Still, the day will hopefully come when society better recognizes the need to reconcile the competing interests of these groups. Hopefully, on that day, policymakers will be well-equipped with promising proposals.

Solar energy and historic preservation

Not surprisingly, very different types of conflicts between renewable energy and cultural preservation tend to arise in dense, urban settings situated great distances from tribal areas. In particular, distributed solar energy installations in urban areas are increasingly running up against historic preservation laws as solar panels appear on more and more rooftops across the globe.

Solar panels vs. historic streetscapes in Washington's Cleveland Park

The struggle of one Washington, D.C., neighborhood to handle a permit request for a rooftop solar panel installation illustrates how historic preservation efforts can clash with solar energy development. The Cleveland Park Historic District is a picturesque suburban neighborhood roughly five miles northwest of Washington's National Mall. Cleveland Park was named after Grover Cleveland and is built partly on the late U.S. president's former estate.[88] As the district's own historical society describes it, the area features a "cohesive collection of single-family residences, apartment buildings, a vibrant commercial corridor with stores and restaurants, schools, a library, firestation and other amenities,"[89] most of which were constructed in the late nineteenth and early twentieth centuries.

The Cleveland Park Historic District is not as culturally or historically significant as the White House, the Washington Monument, or dozens of other iconic sites within Washington, D.C. Nonetheless, the district has appeared on the National Register of Historic Places since the 1980s. Because of this designation, all proposals for renovations or additions to buildings within the district must undergo a detailed design review and comply with a lengthy set of restrictions.[90] These onerous rules seek to protect the district's historic look and feel, thereby helping to preserve property values as well.[91]

In 2012, a controversy arose in Cleveland Park when a pair of neighborhood residents, Mark Chandler and Laurie Wingate, decided they would like to install a solar array on the rooftop of their century-old home. Washington is in the planet's northern hemisphere, so the most productive potential location for a solar array on their home was the building's sloped, south-facing roof, which tended to receive more direct sunlight throughout the year. Panels installed on that part of the roof could have supplied roughly 70 percent of the home's electricity needs.[92]

Unfortunately, Chandler and Wingate knew that putting panels on that area of the roof would not be an option because the home's south side faced a public roadway and was thus off-limits for solar panels. Under the District of Columbia Historic Preservation Guidelines:

> Owners sometimes consider adding solar panels as part of an overall energy efficiency plan for their building. If installed on a flat roof, solar panels should be located so they are not visible from the public street. If located on a sloping roof building, they should only be installed on rear slopes that are not visible from a public street.[93]

In this case, any panels on the couple's southern roof would have been "visible from a public street" and would have violated the guidelines.

Chandler and Wingate also knew that complying with the guidelines by installing panels solely on the back, north-facing side of the roof was not a viable option. An array consisting of panels installed only on the home's back roof would have had far less access to direct sunlight and would have thus generated less than half as much electric power.[94]

Determined to find a way to make a solar energy installation work on their home, Chandler and Wingate approached historic preservation officials and proposed a compromise: an array that occupied portions of the northern and western roofs of the house. Under its proposal, the couple also offered to replace their roof's light-colored shingles with darker shingles to help camouflage the panels.[95] Although some of the panels proposed for their west roof would have been partially visible from the street, any other layout providing less sunlight exposure simply would not have been cost-justifiable. The array needed to be capable of producing enough electricity to pay for itself through energy bill savings within a reasonable number of years.

The stakeholders who seemed most likely to suffer adverse impacts from the couple's proposed solar array expressed vocal support for its installation. The Cleveland Park Historical Society's own Architectural Review Committee indicated that it "support[ed] the installation of solar panels on this property,"[96] noting that the committee had "received very strong statements of support from neighbors adjoining the property."[97] The neighborhood's local "Advisory neighborhood Commission 3C" even passed a resolution in favor of allowing the solar array.[98]

In spite of this neighborhood support, the District of Columbia Historic Preservation Review Board proved unwilling to approve installation of the panels. The board's initial vote on the issue was a 4–4 even split. One board member then abstained and the group voted a second time, voting 4–3 to disallow the new solar panels.

The Board's decision not to approve the proposal seemed to be based on a strict interpretation of the Board's guidelines prohibiting any solar energy installations that were visible from a public street.[99] One board member complained that the panels would "create a visual intrusion on the house and into an intact historic streetscape" and would be "visually distracting from the form and finish of the house's roof."[100] Regardless of the reason, in this particular instance of conflict between solar energy and historic preservation, the preservationists prevailed. Chandler and Wingate ultimately opted not to install solar panels on their home.

Should historic preservation trump solar energy?

A plausible argument could be made that solar panels simply don't belong on the roofs of historic buildings like the home where Chandler and Wingate lived. Very few solar energy systems are even remotely historic in nature or appearance. Indeed, sleek, shiny new solar arrays epitomize modernity. When solar panels find their way into historic neighborhoods or onto the top of vintage buildings, they almost inevitably look out of place. But is that a good enough reason to strictly forbid owners of historic properties from installing solar panels? If not, what policy strategies are best equipped to encourage a fair and efficient balance between solar energy and historic preservation?

Undoubtedly, some buildings do exist that are simply not well suited for solar panel installations.[101] Few people would advocate covering Egypt's iconic pyramids or the famous domes of India's Taj Mahal with solar panels. Preserving our most culturally significant buildings in their original condition and form ensures that they are available to educate and inspire future generations. It would make little sense to blemish the world's most valuable historic resources for just a few kilowatts of new solar generating capacity.

One reason that historic preservation must sometimes take precedence over solar energy development is that solar energy resources are not as location-constrained as most historic sites. Solar radiation strikes the earth with roughly the same intensity on every exposed inch of land within a given community. The sunlight shining onto historic buildings or neighborhoods generally has no greater energy-generating potential than the sunlight shining onto ordinary warehouse rooftops or vacant lots elsewhere in the same town. In contrast, historic properties tend to be most valuable at the precise location where they have always been, and locating them elsewhere is usually not a viable option. Because solar panels can serve their function almost anywhere within a community but historic sites cannot, there is seldom a compelling reason to allow solar energy development to materially interfere with historic preservation. For this reason, laws in some jurisdictions specifically allow local governments to restrict or forbid solar energy projects within historic districts and place no express limitations on the severity of such restrictions.[102]

On the other hand, historic preservation need not always take precedence over solar energy. It is often possible to install solar panels on historic properties without causing any significant harm to their cultural value. In fact, numerous culturally significant buildings around the world are already topped with solar PV systems. In Rome, solar panels grace the roof of a revered Vatican auditorium.[103] The historic Bradford Cathedral in West Yorkshire, England has its own solar array.[104] Rooftop solar panels supply clean electric power to San Francisco's notorious Alcatraz Island.[105] And President Barack Obama had solar panels reinstalled onto the roof of the White House.[106]

If solar panels can adorn the roof the White House—one of the most historic buildings in the world—why are they prohibited on a relatively unfamiliar home just five miles away in Cleveland Park? The answer lies in differences between the types of buildings involved and in the neighborhoods in which they are situated. The White House is situated on land that is largely surrounded by government-owned parcels. The building itself also has a rooftop that is barely visible from the street. Consequently, installing a solar array on the White House roof creates no significant impact on the building's appearance or on surrounding property values. In contrast, the Chandler and Wingate home described above had a steep, sloping roof with no parapets or other features to help hide a solar energy system. The home was also nestled in a residential neighborhood among privately owned homes whose relatively high property values were based in part on the area's historic character. In this very different context, solar panels that were visible from the street could more easily have jeopardized the historic look and feel of the building involved.[107] As illustrated through this comparison, laws must be tailored to account for the various factors at play in clashes between solar energy development and historic preservation, and to incentivize parties to fairly and efficiently mitigate these conflicts.

The benefits of careful planning

Before installing a new solar array, owners of sites that might have cultural or historic significance would be wise to determine whether their properties have already been designated as historic resources under applicable law. Laws in many countries allow citizens to nominate specific properties or neighborhoods for designation as protected historic sites. For example, in the United States, the National Historic Preservation Act of 1966 created a National Register of Historic Places for this purpose.[108] Similar state/ provincial and local historic preservation laws also exist in many jurisdictions.[109] A parcel registered as historic under any of these laws could be subject to special requirements that greatly impact the design of a potential solar energy installation or whether the installation is permissible at all.[110] As in the Cleveland Park case described earlier, local historic preservation officials often have the final say regarding proposed solar energy projects at historic sites.

Even on designated historic buildings, solar energy installations are often feasible through careful planning and design. In addition to the typical list of strategies for minimizing the aesthetic impacts of solar panel installations, researchers have offered a few particular suggestions for those considering solar energy development on historic sites. Such strategies include installing systems only on new additions of historic buildings when possible and not installing solar panels that obstruct views of a roof's "architecturally significant features."[111] As the market for solar energy products grows, manufacturers are also beginning to design solar panels with camouflaging features that could also help to minimize their visual impacts on historic properties. For instance, some companies now sell solar panels in a wide variety of colors. One company sells "tile red" solar panels—panels that blend in far better than conventional panels on rooftops with Spanish tile.[112]

Of course, many strategies for minimizing a solar array's impacts on a historic site also add to the array's cost or reduce its productivity. For example, installing solar panels in less visible areas may also give them less direct sunlight access. Colored solar panels also tend to cost more than conventional ones.[113] These additional costs can potentially dissuade some would-be solar energy users from going solar at all. In the case of Chandler and Wingate, the Historic Preservation Review Board's requirements would have undercut the economics of their proposed project so much that the couple never installed solar panels on their home.

Recognizing that historic preservation laws could deter solar panel installations, lawmakers in some jurisdictions have tried to craft provisions aimed at mitigating this deterrent effect. For instance, statutes in some U.S. states allow local governments to impose conditions or restrictions on solar energy systems in historic districts only "when an alternative of reasonable comparable cost or convenience is available"[114] or when the local restrictions "do not significantly impair" a system's effectiveness.[115] These laws provide some limited guidance to local governments regarding how to balance historic preservation and solar energy. Unfortunately, they contain ambiguous terms such as "reasonable" and "significantly" that create very little certainty for parties operating under such provisions.

Laws in California provide a bit more clarity regarding the balance between solar energy and historic preservation. Statutory provisions enacted in that state allow local governments to impose only "reasonable restrictions" on solar energy installations, defining "reasonable restrictions" as those that "do not significantly increase the cost of the system or significantly decrease its efficiency or specified performance or allow for an alternative system of comparable cost, efficiency, and energy conservation benefits."[116] California's statute goes on to define a "significant" cost increase as "an amount not to exceed two thousand dollars ($2,000) over the system cost as originally specified and proposed, or a decrease in system efficiency of an amount exceeding 20 percent as originally specified and proposed."[117]

California's approach to this issue is appealing in that it provides greater certainty to all parties involved. In situations where the expected neighborhood impacts from a proposed solar array on an historic building are in excess of $2,000 even after the imposition of "reasonable restrictions," the statute may still overly favor solar energy. Nevertheless, the statute is a relatively reasonable and straightforward way of governing with what can be a very complex and fact-specific problem.

Offshore wind energy and cultural preservation at sea

One other potential clash between renewable energy development and cultural resources involves the pristine ocean areas where developers are increasingly seeking to site offshore wind farms. As the following account describes, indigenous peoples nearly halted a prominent offshore wind energy project in the United States by attaching cultural significance to the ocean areas where it was to be built. Although their challenge ultimately proved unsuccessful, these groups exposed an additional source of risk for those seeking to develop offshore wind energy projects.

As described in Chapter 2, when a developer first proposed building the now-infamous Cape Wind project in the Nantucket Sound off the coast of Massachusetts in 2001, the project seemed destined to become the first commercial offshore wind energy project in the United States. Eight years later, after enduring years of opposition in the permitting process, the project finally received a favorable final environmental impact statement in 2009 and appeared on track to be constructed as planned. Then, in October of that year, two Native American tribes cast new doubt on the project by asking federal officials to place the entire 600-square-mile Nantucket Sound on the Federal Register of Historic Places.[118]

The Aquinnah and Mashpee Wampanoag tribes asserted, among other things, that the Nantucket Sound itself was a "traditional cultural property" and that untarnished views of the Sound were of crucial significance in their "sunrise ceremonies."[119] To the surprise of many involved, the National Park Service agreed, officially determining in January of 2010 that:

> Nantucket Sound is eligible for listing in the National Register as a traditional cultural property and as an historic and archeological property ... that has yielded and has the potential to yield important information about the Native American exploration and settlement of Cape Cod and the Islands.[120]

This finding required that Cape Wind's developer fulfill various additional consultation and other requirements in connection with its permit application, further delaying the development of the project.

Ultimately, Secretary of the Interior Ken Salazar approved the project in spite of the historic preservation issues.[121] However, the tribes responded by filing lawsuits challenging Salazar's decision, delaying the project even more. As of late 2013, it appeared likely that construction work would at last commence on Cape Wind within the next year.[122] Still, historic preservation statutes remain an additional weapon for tribal groups and others seeking to impede or frustrate offshore wind energy development.

Charging forward while preserving the past

At some point in the future, humankind may discover a new, transformative energy technology that renders wind turbines and rooftop solar panels totally obsolete. If society ever reaches that point, today's renewable energy systems themselves will then become mere relics of a past era. In such a world, wind turbines and solar panels could conceivably have historic or cultural value to the future generations who ultimately inherit this planet.

Of course, none of that is relevant right now. Indeed, researchers and policymakers across the world today must continue to search for ways to incorporate renewable energy technologies into the landscape without unnecessarily disrupting its physical reminders of humankind's diverse cultural heritage. Although they have their own distinct views of what constitutes conservation, renewable energy advocates and historic preservationists alike are ultimately focused on protecting the world's precious resources for the enjoyment of future generations. Hopefully, these groups can grow to recognize the commonality in their competing goals and work together in pursuing a more culturally rich, sustainable world.

Notes

1 *See, e.g.*, American Planning Association, *Balancing Solar Energy Use with Potential Competing Interests* at 1 (2012), available at www.planning.org/research/solar/briefingpapers/potentialcompetinginterests.htm (last visited Aug. 19, 2013).
2 For example, there are more than 560 federally recognized Indian tribes in the United States. *See* Heather J. Tanana & John C. Ruple, *Energy Development in Indian Country: Working Within the Realm of Indian Law and Moving Towards Collaboration*, 32 UTAH ENVTL. L. REV. 1, 2 (2012).
3 In the United States, much of this relocation process was initiated under the Indian Removal Act of 1830. For a brief overview of this legislation and its impacts and links to numerous other sources on this topic, visit the United States Library of Congress website at www.loc.gov/rr/program/bib/ourdocs/Indian.html (last visited Dec. 18, 2013).
4 For general information about coal production at the Kayenta Mine in the Black Mesa area and use of the mined coal at the Navajo generating station, visit the Peabody Energy website at www.peabodyenergy.com/content/276/Publications/Fact-Sheets/Kayenta-Mine (last visited Aug. 20, 2013).
5 *See* Douglas C. MacCourt, RENEWABLE ENERGY DEVELOPMENT IN INDIAN COUNTRY: A HANDBOOK FOR TRIBES, NREL & AterWynne, LLP (June 2010) (available at http://apps1.eere.energy.gov/tribalenergy/pdfs/indian_energy_legal_handbook.pdf (last visited Aug. 12, 2013) (citing *DOE Office of Energy Efficiency and Renewable Energy (EERE), DOE's Tribal Energy Program*, PowerPoint Presentation prepared by Lizana K. Pierce (available at http://apps1.eere.energy.gove/tribalenergy/)).
6 *See* id.

7 *See, e.g.*, Sara C. Bronin, *The Promise and Perils of Renewable Energy on Tribal Lands*, 26 TUL. ENVTL. L.J. 221 (2013) (noting that, as of 1990, about 16,000 tribal households in the U.S. "lacked access to electricity or other arrangements that would provide electricity at no cost," a rate that is "ten times the national average"). Bronin ultimately argues in her article that small-scale renewables offer the greatest hope for electrification of remote tribal lands. *See* id. at 235. *See also* Matt Rivera and John Roach, *Out of Darkness: Solar Power Sheds a Little Light on Powerless Communities*, NBC NEWS (Aug. 11, 2013), available at www.nbcnews.com/technology/out-darkness-solar-power-sheds-little-light-powerless-communities-6C10867721#out-darkness-solar-power-sheds-little-light-powerless-communities-6C10867721 (last visited Aug. 12, 2013) (describing the increasing use of solar panels to provide evening light to families in the Navajo nation and in rural areas of Africa).

8 *See* Eve Darian-Smith, *Environmental Law and Native American Law*, 6 ANNU. REV. LAW SOC. SCI. 359, 361 (2010) (noting that it is important "not to treat all native peoples as one homogeneous category or to consider their thinking about nature as consistent across all tribes and individuals").

9 Id.

10 *See* Crystal D. Masterson, *Wind-Energy Ventures in Indian Country: Fashioning a Functional Program*, 34 AMER. IND. L. REV. 317, 337 (2009–10) (noting that "Tribal 'subsistence, culture, and spirituality are intimately connected to the lands they inhabit'") (citing Jacqueline P. Hand, *Global Climate Change: A Serious Threat to Native American Lands and Culture*, 38 ENVTL. L. REP. NEWS & ANALYSIS 10329, 10330 (2008).

11 *See, e.g.*, Denise De Paolo, *Oglala Sioux Make Strides Toward Energy Independence*, KSFY.com (July 15, 2013), available at www.ksfy.com/story/22847509/oglala-sioux-make-strides-toward-energy-independence (last visited Aug. 12, 2013) (quoting member of the Oglala Sioux tribe favoring responsible renewable energy development on tribal lands for its ability to promote the "protection of the land for the future of our people, our generations, our youth").

12 Ryan David Dreveskracht, *Economic Development, Native Nations, and Solar Projects*, 72 AMER. J. ECON. & SOC. 122, 136 (2013).

13 Other scholars have made note of these challenges. *See, e.g.*, Darian-Smith, *supra* note 8 at 337 (arguing that, "[o]n account of their intimate relationship with nature, tribes may be less inclined than non-tribal entities to surrender the pristine landscape of their reservations in exchange for the affluence attainable from the installation of a wind-power project.").

14 Paul E. Frye, *Developing Energy Projects on Federal Lands: Tribal Rights, Roles, Consultation, and Other Interests (a Tribal Perspective)*, ROCKY MOUNTAIN MINERAL LAW FOUNDATION-INST., No. 3, Paper No. 15B (Sept. 2009).

15 *See* Bronin, *supra* note 7 at 234.

16 *See, e.g.*, Kristi Eaton, *Sioux Tribes Plan Large-Scale Wind Energy Project*, YAHOO! NEWS (Associated Press) (July 5, 2013), available at http://news.yahoo.com/sioux-tribes-plan-large-scale-175702015.html (last visited Aug. 14, 2013) (noting that a proposed major wind energy project on Sioux reservation lands in North Dakota would require "several major [new transmission] lines, which cost about $1 million per mile and take up to a year and a half to build").

17 25 U.S.C. § 3502 (2006).

18 *See* Bronin, *supra* note 7 at 229 (describing the Division of Indian Energy Policy Development and the Office of Indian Energy Policy and Programs).

19 *See generally* Energy Policy Act § 502 et seq., Public Law 109–58 (2005).
20 These proposed amendments, titled the Indian Energy Promotion and Parity Act of 2010, sought to amend the Energy Policy Act of 1992 and sought to provide assistance and simplified approval process for American Indian tribes seeking to do energy development on their lands. Despite multiple attempts, these amendments had not yet been enacted as of 2013. For a discussion of the amendments, *see generally* Kelly de la Torre & Robert S. Thompson III, *The Indian Energy Promotion and Parity Act of 2010: Opportunities for Renewable Energy Projects in Indian Country,* 2011 No. 2 RMMLF-INST Paper No. 8 (2011).
21 *See, e.g.,* Ryan D. Dreveskracht, *Alternative Energy in American Indian Country: Catering to Both Sides of the Coin,* 33 ENERGY L.J. 431, 440–48 (2012); de la Torre & Thompson, *supra* note 20.
22 *See* David Agnew, *Responsibility to Future Generations: Renewable Energy on Tribal Lands,* www.whitehouse.gov (June 15, 2013), available at www. whitehouse.gov/blog/2013/06/25/responsibility-future-generations-renewable -energy-development-tribal-lands (last visited Aug. 12, 2013).
23 *See* id.
24 Heather Beck, *Cooktown One Step Closer to Wind Farm,* CAIRNS POST (May 12, 2011), available at www.cairns.com.au/article/2011/05/12/163591_local-news.html (last visited Aug. 12, 2013).
25 *See* Siobhan Barry, *Indigenous Backing for Cooktown Wind Project,* ABC NEWS (Australia) (Aug. 11, 2009), available at www.abc.net.au/news/2009-08-12/indigenous-backing-for-cooktown-wind-project/1387704 (last visited Aug. 12, 2013).
26 *See* United States Department of Interior Instruction Memorandum No. 2011-061 (Feb. 7, 2011), available at www.blm.gov/pgdata/etc/medialib/blm/wo/Communications_Directorate/public_affairs/news_release_attachments. Par.79538.File.tmp/IM2011.61.Prescreening.pdf (last visited Aug. 13, 2013).
27 Some geothermal energy projects have been under consideration as well. To see an updated list of these projects and their status, visit the BLM website at www.blm.gov/wo/st/en/prog/energy/renewable_energy/Renewable_Energy_Projects_Approved_to_Date.html (last visited Aug. 13, 2013).
28 Bureau of Land Mgmt., *Solar Energy Development Draft Programmatic Environmental Impact Statement 5-218* (2010) (cited in John Copeland Nagle, *Green Harms of Green Projects,* 27 NOTRE DAME J. OF LAW, ETHICS & PUB. POL'Y 59, 72 (2013)).
29 Bureau of Land Mgmt., *Solar Energy Development Final Programmatic Environmental Impact Statement 5-21* (2012) (cited in Nagle, *supra* note 28 at 72).
30 Bureau of Land Mgmt., *Final Programmatic Environmental Impact Statement on Wind Energy Development in Six Southwestern States 5-99* (2012) (cited in Nagle, *supra* note 28 at 72).
31 National Historic Preservation Act § 101(d)(6)(B), 16 U.S.C. § 470a (2006).
32 *See* 16 U.S.C. § 470f (2006); 36 C.F.R. § 800.2-800.6.
33 *See Quechan Tribe of the Fort Yuma Indian Reservation v. U.S. Dep't of the Interior,* 755 F.Supp.2d 1104, 1106 (2010).
34 Id. at 1107.
35 Id.
36 Id. at 1119.
37 Id.
38 Id.
39 *See* Nagle, *supra* note 28 at 82.

40 *See* id.
41 See Morgan Lee, *Ocotillo Wind Project Advances Despite Tribal Objections*, SAN DIEGO UNION-TRIBUNE (May 12, 2012).
42 Complaint of Quechan Indian Tribe for Declaratory and Injunctive Relief at ¶ 5, *Quechan Tribe of the Fort Yuma Indian Reservation v. U.S. Dep't of the Interior* (S.D. Cal. May 14, 2012).
43 *Quechan Tribe of the Fort Yuma Indian Reservation v. U.S. Dep't of the Interior*, 2013 WL 755606 at 1 (S.D. Cal. Feb. 27, 2013).
44 Id.
45 Miriam Raftery, *Native American Heritage Commission Declares Ocotillo Wind Farm Site a Sacred Site; Asks Attorney General to Weigh Legal Action*, EAST COUNTY MAGAZINE (Apr. 26, 2013), available at http://eastcountymagazine.org/node/13103 (last visited Aug. 14, 2013).
46 *See, e.g.*, Jessica Testa, *Sacred Ground? Citing 'Viewshed,' Tribe Pushes Back Against Solar Plant*, CRONKITE NEWS (May 1, 2012), available at http://cronkitenewsonline.com/2012/05/sacred-ground-citing-significant-views-tribe-pushes-back-against-solar-plant/ (last visited Aug. 15, 2013).
47 *See generally* Louis Sahagun, *Discovery of Indian Artifacts Complicates Genesis Solar Project*, LOS ANGELES TIMES (April 24, 2012).
48 Id.
49 *See* Letter re: Comments of the Colorado River Indian Tribes on the DRECP Interim Document (Docket No. 09-RENEW EO-01) from the Colorado River Indian Tribes to the California Energy Commission (Jan. 30, 2013), available at www.drecp.org/documents/docs/comments-evals/Colorado_River_Indian_Tribes_comments.pdf (last visited Aug. 15, 2013).
50 Id. at 5 (citing California Energy Commission Desert Renewable Energy Conservation Plan Interim Document (Docket No. 09-RENEW EO-01), 4.3-4 (2012)).
51 *See* Indigenous Peoples Subcomm., Nat'l Envtl. Justice Advisory Council, EPA, *Guide on Consultation and Collaboration with Indian Tribal Governments and Public Participation of Indigenous Groups and Tribal Members in Environmental Decision Making* at 5 (EPA/300-R-00-009, 2000), available at www.epa.gov/environmentaljustice/resources/publications/nejac/ips-consultation-guide.pdf (last visited Aug. 15, 2013)
52 Id. at 14.
53 Quechan Tribe, *supra* note 33, 55 F. Supp.2d at 1119.
54 Lee, *supra* note 41.
55 For example, in the Canadian province of Canada, disputes have arisen between the Ojibwe tribe and the city of Thunder Bay over a proposed 16-turbine wind farm on city-owned land in the Nor'Wester Mountains. According to a newspaper article, the Ojibwe have "lived and worshipped" on the project site for "thousands of years" and have always considered the mountain range as a "sacred place." *See Ontario Ojibwe Rally to Battle Wind-Farm Plans on Sacred Nor'Wester Mountains*, INDIAN COUNTRY TODAY MEDIA NETWORK (May 29, 2013), available at http://indiancountrytodaymedianetwork.com/2013/05/29/ontario-ojibwe-rally-battle-wind-farm-plans-sacred-norwester-mountains-149599 (last visited Aug. 16, 2013).
56 Marcelino Madrigal & Steven Stoft, *Transmission Expansion for Renewable Energy Scale-Up: Emerging Lessons and Recommendations* at 95, The Energy and Mining Sector Board (June 2011), available at http://web.worldbank.org/WBSITE/EXTERNAL/TOPICS/EXTENERGY2/0, contentMDK:22990540~pagePK:210058~piPK:210062~theSitePK:4114200,00.html (last visited Feb. 24, 2014).

57 Rachael Peterson, *Mexico: Federal Court Halts Controversial Wind Park*, Globalvoicesonline.com (Dec. 27, 2012), available at http://globalvoicesonline. org/2012/12/27/mexico-federal-court-halts-controversial-wind-park/ (last visited Aug. 15, 2013).
58 David LaGesse, *Mexico's Robust Wind Energy Prospects Ruffle Nearby Villages in Oaxaca*, NATIONAL GEOGRAPHIC-DAILY NEWS (Feb. 7, 2013), available at http://news.nationalgeographic.com/news/energy/2013/02/pictures/130208- mexico-wind-energy/ (last visited Aug. 15, 2013).
59 *See* Peterson, *supra* note 57.
60 *See* LaGesse, *supra* note 58.
61 *See* Emilio Godoy, *Rural Mexican Communities Protest Wind Farms*, INTER PRESS SERVICE (June 19, 2013), available at www.ipsnews.net/2013/06/rural- mexican-communities-protest-wind-farms/ (last visited Aug. 15, 2013).
62 *See* Peterson, *supra* note 57.
63 As of August 2013, one Mexican peso was valued at less than $.08 in U.S. dollars. *See* http://themoneyconverter.com/MXN/USD.aspx (last visited Aug. 15, 2013).
64 *See* Emilio Godoy, *supra* note 61.
65 *See* Peterson, *supra* note 57.
66 International Development Exchange, *Oaxaca Wind Farm Urgent Action* (Feb. 8, 2013), available at https://www.idex.org/oaxaca-wind-farm-urgent-action/ (last visited Feb. 12, 2014).
67 Jennifer M. Smith, *Indigenous Communities in Mexico Fight Corporate Wind Farms*, upsidedownworld.org (Nov. 1, 2012), available at http://upside- downworld.org/main/mexico-archives-79/3952-indigenous-communities -in-mexico-fight-corporate-wind-farms (last visited Aug. 16, 2013).
68 *See* id.
69 Id.
70 *See* id.
71 *See* Peterson, *supra* note 57 (citing English version of Rosa Rojas, Mexico's Indigenous Appeal Wind Project, LA JORNADA (June 13, 2013).
72 *See* International Development Exchange, *supra* note 66.
73 *See* id.
74 *See* Michael McGovern, *Land Right Protests Hold Up Construction in Oaxaca*, windpowermonthly.com (Dec. 1, 2012), available at www. windpowermonthly.com/article/1161708/land-right-protests-hold- construction-oaxaca (last visited Aug. 16, 2013); *see also* Erik Vance, *The "Wind Rush": Wind Energy Blows Trouble in Mexico*, CHRISTIAN SCIENCE MONITOR (Jan. 26, 2013), available at www.csmonitor.com/ Environment/2012/0126/The-wind-rush-Green-energy-blows-trouble-into- Mexico (last visited Nov. 7, 2013).
75 *Oaxaca: Communards from Unión Hidalgo Request That the TUA Totally Nullify Demex Contracts*, SIPAC Blog (June 24, 2013), available at http:// sipazen.wordpress.com/tag/zapotec/ (last visited Aug. 16, 2013).
76 Id.
77 *See, e.g.*, Charlotte Shipman, *Maori Water Rights Claim at Supreme Court*, 3NEWS (New Zealand) (Sept. 5, 2012), available at www.3news.co.nz/Maori- water-rights-claim-at-Supreme-Court/tabid/1607/articleID/285150/Default. aspx (last visited Aug. 16, 2013) (noting that "[t]he Crown accepts Maori have some rights over water—but not full ownership").
78 Shelley Bridgeman, *Who Owns the Wind?*, NEW ZEALAND HERALD (Sept. 13, 2012), available at www.nzherald.co.nz/lifestyle/news/article.cfm?c_ id=6&objectid=10833580 (last visited Aug. 16, 2013).

79 Dan Satherley, *Treaty Gives Us the Wind—Ngapuhileader*, 3NEWS (New Zealand) (Sept. 5, 2012), available at www.3news.co.nz/Treaty-gives-us-the-wind---Ngapuhi-leader/tabid/1607/articleID/268086/Default.aspx (last visited Aug. 12, 2013).

80 *See* Bridgeman, *supra* note 78 (noting that the Maori "hope[d] to earn dividends for wind used in the commercial generation of electricity).

81 *See* id. (Stating that, "[a]fter wind, Mr Rankin believes Maori will look to claim other property rights, giving aerospace as an example").

82 *See generally* Rawiri Smith, *A Cultural Impact Assessment of Genesis Energy's Castle Hill Wind Farm Before the AEE* (July 21, 2011), available at http:// dev.horizons.govt.nz/assets/Volume-4bSection-4-Cultural-Impact-and-Values-Effects-Assessments-3.-Ngati-Kahungunu-Ki-Wairarapa.pdf (last visited Feb. 24, 2014).

83 Id. at 18–19.

84 Id. at 31.

85 *See* KIWI (NZ) TO ENGLISH DICTIONARY, available online at www.newzealan-datoz.com/index.php?pageid=357 (last visited Aug. 16, 2013).

86 *See* Business Desk, *Genesis Gains Consent for Wairarapa Mega-wind Farm*, THE NATIONAL BUSINESS REVIEW (Jul. 4, 2013), available at www.nbr.co.nz/article/genesis-gains-consent-wairarapa-mega-wind-farm-bd-142453 (last visited Aug. 16, 2013).

87 A member of the Navajo nation suggested the possibility of such an Indian carve-out in RPS standards in hearings about proposed amendments to the ITEDSA in 2012. *See* S. Hrg. 112–636 at 43 (April 19, 2012).

88 *See* Kimberly Prothro Williams, *Cleveland Park Historic District* (Brochure) at 1 (2001), available at http://planning.dc.gov/DC/Planning/Historic+Preservation/Maps+and+Information/Landmarks+and+Districts/Historic+District+Brochures/Cleveland+Park+Historic+District+Brochure (last visited Aug. 18, 2013).

89 *See* id.

90 For general information about the Washington, D.C., Historic Preservation Law (D.C. Law § 2-144), and how it applies to Cleveland Park, visit the historic district's website at www.clevelandparkhistoricalsociety.org/historic-district/doing-work-on-your-historic-district-home/ (last visited Aug. 19, 2013).

91 *See* Kaid Benfield, *Can Solar Panels and Historic Preservation Get Along?*, THE ATLANTIC CITIES (June 25, 2012), available at http://www.citylab.com/cityfixer/2012/06/can-solar-panels-and-historic-preservation-get-along/2364/ (last visited Jun. 3, 2014) (referring to research showing that "architecturally inappropriate changes" to homes in historic districts can cause "property value reduction[s] of around $60,000 for abutting and nearby properties").

92 *See* Kathy Orton, *Solar Power Project Eclipsed in D.C.*, WASHINGTON POST (June 21, 2012), available at http://articles.washingtonpost.com/2012-06-21/news/35461157_1_solar-panels-roof-solar-power (last visited Aug. 19, 2013).

93 *See* District of Columbia Historic Preservation Office, *District of Columbia Historic Preservation Guidelines: Roofs on Historic Buildings* at 12 (2010), available at http://dc.gov/OP/HP/Guideline%20pdf%20files/DC_Roof_Guidelines _SW.pdf (last visited Aug. 19, 2013).

94 *See* Orton *supra* note 92 (noting that putting panels on the rear of the home would have allowed the array to fill only about 30 percent of the home's electricity needs).

95 *See* Benfield, *supra* note 91.

96 David Alpert, *Preservation Staff Reject Solar Panels on Cleveland Park Home*, GREATER GREATER WASHINGTON (May 31, 2012), available at http://

greatergreaterwashington.org/post/15012/preservation-staff-reject-solar-panels-on-cleveland-park-home/ (last visited Aug. 19, 2013).

97 Id.

98 *See* Orton, *supra* note 92.

99 *See* id.

100 Id.

101 At least one researcher of this topic has made a similar observation. *See* Kimberly Kooles, et al., *Installing Solar Panels on Historic Buildings*, U.S. Dept. of Energy (August 2012), available at http://ncsc.ncsu.edu/wp-content/uploads/Installing-Solar-Panels-on-Historic-Buildings_FINAL_2012.pdf   (last visited Aug. 19, 2013) (noting that "there will be certain historic properties for which solar energy systems may not be appropriate").

102 *See, e.g.*, N.M. Stat. § 3-18-32 (specifically allowing restrictions on solar energy systems in historic districts).

103 *See* Frances d'Emilio, *Vatican Pushes Solar Energy*, Boston Globe (June 6, 2007).

104 *See* Jim Greenhalf, *Solar Panels Go on Bradford Cathedral Roof*, Telegraph & Argus (Aug. 24, 2011), available at www.thetelegraphandargus.co.uk/news/9213579.Solar_panels_go_on_Bradford_Cathedral_roof/ (last visited Aug. 19, 2013).

105 *See* Justin Gerdes, *Solar-Powered Microgrid Slashes Alcatraz Island's Dependence on Fossil Fuels*, Forbes (July 30, 2012), available at www.forbes.com/sites/justingerdes/2012/07/30/solar-powered-microgrid-reduces-alcatraz-islands-dependence-on-fossil-fuels/ (last visited Aug. 19, 2013).

106 Jimmy Carter was the first president to install solar panels on the White House roof, having done so during his presidency in the late 1970s. *See* Juliet Eilperin, *White House Solar Panels Being Installed This Week*, Washington Post (Aug. 15, 2013), available at www.washingtonpost.com/blogs/post-politics/wp/2013/08/15/white-house-solar-panels-finally-being-installed/ (last visited Aug. 19, 2013).

107 *See* note 91, *supra*.

108 *See generally* 16 U.S.C. § 470 (1992).

109 For a list of U.S. state and local historic preservation laws relating to solar energy installations, *see* A. Kandt, et al., *Implementing Solar PV Projects on Historic Buildings and in Historic Districts* at 21–23, NREL/TP-7A40-5197 (Sept. 2011), available at www.nrel.gov/docs/fy11osti/51297.pdf (last visited Aug. 20, 2013).

110 For a detailed description of U.S. historic preservation laws and how they might affect distributed solar energy development, *see* id. at 3–9.

111 *See* id. at 20.

112 *See* Emily Hois, *Colored Solar Panels Address Concerns of Aesthetics, Historic Preservation*, solarreviews.com (April 30, 2013), available at www.solarreviews.com/blog/colored-solar-panels-address-concerns-of-aesthetics-historic-preservation/ (last visited Aug. 20, 2013).

113 *See, e.g.*, Hois, *supra* note 112 (noting that "consumers should expect to pay two or three times more for adding color to normal PV cells").

114 Maine Rev. Stat. Ann. § 1423.4.C.

115 Conn. Gen. Stat. § 7-147(a).

116 Cal. Civ. Code § 714(b).

117 Cal. Civ. Code § 714(d)(1)(B).

118 *See* Timothy H. Powell, *Revisiting Federalism Concerns in the Offshore Wind Energy Industry in Light of Continued Local Opposition to the Cape Wind Project*, 92 B.U. L. Rev. 2023, 2039 (2012).

119 *See* Kenneth Kimmell & Dawn Stolfi Stalenhoef, *The Cape Wind Offshore Wind Energy Project: A Case Study of the Difficult Transition to Renewable Energy*, 5 GOLDEN GATE U. ENVTL. L.J. 197, 209 (2011).
120 Nat'l Park Serv., U.S. Dep't of the Interior, *National Register of Historic Places Determination of Eligibility Comment Sheet 4* (2010), available at www.nps. gov/nr/publications/guidance/NantucketSoundDOE.pdf (last visited Oct. 18, 2013) (cited in Powell, *supra* note 118 at 2039–40).
121 Powell, *supra* note 118 at 2040.
122 Dave Levitan, *Will Offshore Wind Finally Take Off on U.S. East Coast?*, YALE ENVIRONMENT 360 (Sept. 23, 2013), available at http://e360.yale.edu/feature/will_offshore_wind_finally_take_off_on_us_east_coast/2693/ (last visited Oct. 16, 2013).

7 Renewable energy vs. the electric grid

As if opposition from neighbors, special interest groups, and competing developers were not enough, challenges associated with the electric grid are increasingly hindering renewable energy development as well. Even the most promising wind and solar energy project sites are undevelopable without a feasible way of delivering their generated power to end users. For a variety of frustrating reasons, the existing electric transmission infrastructure in many countries is all too often incapable of supporting new wind and solar energy projects in regions with the best renewable energy resources. Outmoded features of electric grids can also limit their ability to accommodate rooftop solar energy systems, small wind turbines, and other forms of distributed generation. Unless industry stakeholders find ways around these challenges, grid constraints will continue to be a major roadblock to the sustainable energy movement.

The grid-related conflicts that can slow growth in the renewable energy sector come in three basic varieties. A first category of conflicts involves *transmission* issues—difficulties relating to the delivery of electricity from remote regions where large wind and solar energy projects tend to be sited to metropolitan areas interested in purchasing that power. These transmission obstacles exist because of the geographic realities of utility-scale renewable energy development. Many of the world's most productive renewable energy resources are in locations that are distant from major population centers and the electricity infrastructure that supports those areas. Expensive high-voltage transmission lines are the only practical way to transmit power from these remote wind and solar farm sites to consumers in bustling cities. Disputes over who will fund the massive transmission expansion projects necessary to support renewable energy are increasingly pitting governments against each other, often resulting in standstills and inadequate levels of investment. Thick layers of regulatory red tape and disagreements among governments about where new transmission lines should be sited only exacerbate these problems.

A second category of grid-related conflicts involving renewable energy consists of *distribution* issues—controversies resulting from the disruptive impact that wind and solar energy systems can have on the day-to-day operation and maintenance of electric grids. In contrast to *transmission*, which has been defined as the "movement of electricity over longer distance at higher voltages,"[1] *distribution* is the "delivery of electricity at lower voltages from high-voltage transmission lines to end users."[2] Wind and sunlight are intermittent energy resources—at any given location, they vary in amount from hour to hour or even from second to second. When wind and solar energy projects feed power onto a grid system, the variable nature of these resources can complicate grid operators' efforts to balance electricity supply and demand and increase the risk of disruptive blackouts or of damaging grid overloads.

A third category of grid-related conflicts are disputes involving utilities that have come to view renewable energy as a direct threat to their long-held position within the electricity industry. In some markets, prices for wind-generated electricity have dropped to such low levels that generators of non-renewable power are having difficulty finding buyers during certain times of the day. The growth of rooftop solar energy and "net metering" has likewise begun to affect utilities' bottom lines and market shares in some jurisdictions. These trends have led some utilities to begin seeking to impede renewable energy development as a strategy for protecting their very survival.

This chapter examines how each of these three categories of grid-related obstacles is hindering renewable energy growth around the world. The chapter also discusses some potential means of overcoming these challenges, or of at least preventing them from hindering the global push toward a more sustainable energy future.

Conflicts over expanding transmission systems to accommodate renewable energy

Utility-scale wind and solar energy development is placing unprecedented pressure on aging electric transmission systems across the planet. These systems were largely designed and constructed in a different energy era— one dominated by traditional coal-, nuclear-, or gas-fired power plants whose energy sources were typically delivered directly to plants via pipeline, truck or rail. Because of the nature of these conventional energy sources, officials historically had a fair amount of flexibility when siting traditional power plants and could put them in locations that best suited a region's or community's energy needs.

Of course, siting wind farms and large solar energy plants inherently involves far less flexibility than the siting of conventional power plants. Unlike fossil fuels, wind currents and sunlight are not transportable by rail or pipeline and generally must be converted into electricity instantane-ously in the precise locations where they naturally occur.[3] The quality of wind and solar resources can also vary dramatically across continents and regions. Because of these immutable characteristics of wind and sunlight, officials tend to have far less leeway in siting utility-scale solar plants and wind farms. Sites that lack strong natural wind or solar resources are not even worth considering for such projects, even if they would otherwise be convenient based on their proximity to cities or existing transmission corridors.

A large proportion of the world's most productive wind and solar energy resources are found in remote areas that are hundreds of miles away from major urban centers. Consequently, major transmission upgrades or expansions are often necessary to connect new utility-scale renewable energy projects to the grid. Existing electric transmission systems in most of the world were primarily designed to distribute power from conventional power plants to nearby end-users. These decades-old grids are often incapable of delivering renewable power from the far-flung places where it can be commercially harvested to electricity consumers. Due to these transmission limitations, thousands of would-be wind farm and solar farm sites around the globe today are little more than isolated islands of untapped renewable energy sources.

The transmission challenges affecting wind energy developers in the western United States exemplify how deficiencies in transmission infrastructure can slow the pace of renewable energy development. As shown in Table 7.1 below, the U.S. states with the best onshore wind energy resources generally are situated in inland western and mid-western regions of the country—"flyover states" where cities are relatively few and far between. Of the top ten states for wind energy potential, only Texas has a total population that ranks in the top twenty out of fifty states. Because of relatively low population density in most of the United States' interior west, the electric transmission infrastructure in this region has historically been limited. These aging transmission facilities were largely designed based on the assumption that a mix of hydropower and locally-sited conventional power plants would continue to supply the region's energy for the

Table 7.1 The geographic mismatch between wind energy resources and population density in the United States

Top ten U.S. states for wind energy resources[4]	Population rank in 2010 (out of 50 states)[5]
1. North Dakota	48
2. Texas	2
3. Kansas	33
4. South Dakota	46
5. Montana	44
6. Nebraska	38
7. Wyoming	50
8. Oklahoma	28
9. Minnesota	21
10. Iowa	30

foreseeable future. As more and more wind energy projects have connected onto the grid in these states, grid operators have struggled to find new ways to address the congestion problems introduced by the addition of this new power source.[6]

An obvious solution to the sorts of challenges just described would be to upgrade and expand existing transmission systems to accommodate the new influx of wind and solar energy projects. Unfortunately, very few governments are successfully doing that, and renewable energy projects are languishing as a result. In California, there are "many projects in the planning stages yet unable to be realized because of grid constraints."[7] And in the United Kingdom, officials have informed some renewable energy developers that grid access would be unavailable until as late as 2025 because of limitations in that nation's outmoded web of transmission lines.[8]

One downside of any major transmission build-out for renewable energy is its potential to blemish pristine landscapes, disturb landowners, and disrupt sensitive habitat areas. Above-ground, high-voltage transmission lines are not renowned for their beauty. There is also some limited evidence suggesting that the electromagnetic fields associated with high-voltage transmission facilities may pose a health hazard to humans.[9] It is thus hardly surprising that contingent valuation studies have found that the perceived aesthetic blight and annoyance caused by the presence of high-voltage lines in a community can erode local property values.[10]

Installing hundreds of miles of massive new transmission facilities and clearing of wide pathways through forested areas to accommodate them can pose threats to birds and other wildlife species and attract opposition for those reasons as well.[11] Bird electrocutions, habitat disruptions, and countless other adverse impacts can result when a new high-voltage transmission line runs through an otherwise pristine rural area. Concerned about these and other impacts, landowners and local governments can be formidable opponents of high-voltage transmission development.

In light of the potential problems associated with massive build-outs of transmission grids, some scholars have advocated limiting these conflicts by relying more on distributed generation and "microgrids"—systems that allow for more localized, decentralized generation and distribution of electric power.[12] For example, commercial-scale solar PV installations on the tops of warehouses, superstores, and other large buildings within urban areas can be a relatively low-cost means of adding additional renewable energy to a region's electricity mix without the need for lots of new, long-distance transmission lines.[13] On average, utility-scale solar PV plants situated in remote desert areas are likely to have greater adverse ecological and other impacts than discretely-positioned rooftop solar arrays inside a city. Generating "distributed" renewable energy through rooftop solar arrays and other small- to medium-scale renewable energy systems also avoids the energy losses associated with transmitting power across long distances—up to 0.7 percent per 100 miles of 500 kV line.[14] Recognizing the benefits of distributed generation, Germany has emerged as a leader in this green energy strategy, installing thousands of megawatts of new solar PV generating capacity per year in the form of relatively small projects.[15]

Although distributed generation through smaller-scale wind and solar energy systems is an increasingly valuable part of the global sustainable energy movement, it also has its limitations. For instance, distributed generation strategies are incapable of making use of the large quantity of valuable wind and solar energy resources available in rural areas. Without careful siting and design, renewable energy systems installed in urban environments are also more likely to trigger disputes over such issues as aesthetics and solar access. And, as described later in this chapter, increases in distributed generation can complicate utility operators' task of balancing the supply and demand of power on the grid. Because of these and other constraints, distributed renewable energy is unlikely to be a panacea for the transmission-related problems associated with renewable energy. For the planet to fully utilize its wind and solar energy resources, a substantial proportion of the world's renewable power will likely need to travel across long distances before reaching consumers.

Recognizing that dramatic improvements in high-voltage transmission capacity are essential to the continued growth of its business, the wind energy industry in the United States has advocated for the development of an expansive new "backbone" system of extra high-voltage (EHV) lines across that country's Midwest region.[16] This enormous system would be capable of delivering large amounts of wind-generated power from the windswept hills and prairies of America's flyover country to population centers further east and along the Pacific coast. Such a system, if constructed, would significantly reduce grid congestion problems and would provide enough transmission capacity to enable rapid wind energy development to continue in the region for several decades.

To date, the U.S. wind energy industry's vision of a transmission super-highway to spread wind-generated electricity throughout the continent has yet to be realized, and transmission constraints continue to hamper renewable energy growth in the U.S. and throughout the world. Why does this lack of adequate transmission infrastructure continue to be such a troublesome barrier to the global transition to renewable energy resources? Although the factors contributing to this problem vary from country to country, they often have something to do with the regulatory regimes that govern the siting and funding of transmission facilities.

The problem of multiple siting authorities

In some parts of the world, the sheer number of government entities capable of blocking any given transmission project proposal has stifled efforts to update and expand grids for renewable energy development. Because major transmission projects often stretch across multiple state or national boundaries, they can be subject to the regulatory authority of each of the many jurisdictions into which they cross. The absence of a "one-stop shop" capable of offering all necessary approvals for an entire cross-jurisdictional transmission line project means that any one of the several agencies having regulatory power over some part of that proposed line is capable of single-handedly blocking the project.[17] This fragmented, decentralized regulatory environment can itself impede the siting of transmission facilities for renewable energy. As one pair of energy scholars describes this problem:

> The more regulatory bodies to which an applicant must apply, the higher the process costs will be and the more likely it is that the litigiousness of intervenors could drive those costs even higher. The result could well be to discourage investors from committing to projects and applicants or potential applicants from building.[18]

The fragmented nature of the transmission line approval process in many countries complicates development efforts because almost any such project is bound to create some winners and some losers. Even the most socially valuable interjurisdictional transmission projects tend not to equally benefit or burden all of the jurisdictions involved. Some jurisdictions may be in favor of a proposed transmission project for the additional electricity it could bring within their boundaries or for its potential to facilitate additional local renewable energy development and the new jobs and tax revenue that such development could provide. In contrast, other jurisdictions that would host portions of a proposed line may conclude that its adverse impacts on pristine landscapes or property values would result in net losses within their boundaries. Jurisdictions in these latter situations have incentives to exercise whatever authority they possess to prevent the project from moving forward.

The disadvantages of having too many regulatory authorities involved in the long-distance transmission process have been evident in connection with multiple proposed transmission expansions in the United States. Consider, for instance, the relatively recent conflict over a proposed 230-mile high-voltage transmission line that would have stretched from Palo Verde, Arizona, to Devers, California.[19] The line, formally proposed in 2007, would have enabled more electric power generated in Arizona to be sold to end-users in California. Because Southern California Edison, a California utility, had offered to build the line, its real-dollar cost would have been borne by California ratepayers and not by Arizonans.[20] Still, the project needed approval from regulators in California and in Arizona. Although California regulators enthusiastically granted their approval for the line, Arizona regulators refused to approve it. They expressed concern about the line's potential environmental impacts on the state and suggested that it would be nothing more than a "giant extension cord" designed to deliver Arizona electricity to California consumers.[21] Able to secure a thumbs-up from regulators in only one of the two states involved in the project, Southern California Edison never built the line as planned.

Federal- and state-level attempts to address the problem

In the United States, these sorts of challenges are attributable in part to federal laws that allow individual states to retain significant control over the siting of any new transmission facilities within their geographic boundaries. At the federal government level, the primary U.S. agency that oversees interstate transmission development is the Federal Energy Regulatory Commission (FERC). Under the Federal Power Act, FERC possesses authority to regulate "the transmission of electricity in inter-state commerce" within the United States.[22] However, that authority extends "only to those matters which are not subject to regulation by the States."[23] Based on this language, individual states have historically held independent approval rights for interstate transmission projects. When a transmission line is proposed that would cross through multiple states, each state in the line's path has its own separate authority to prevent the project from proceeding. Such disjointed governance can be a developer's worst nightmare: in one instance, a U.S. company reportedly spent 14 years collecting approvals for an interstate transmission line that ultimately took just 18 months to physically build.[24]

In an effort to address these issues, the U.S. Congress enacted some legislative changes in the Energy Policy Act of 2005 (EPAct 2005) aimed at streamlining and simplifying the approval process for interstate transmission lines. Among other things, the EPAct 2005 contained provisions purporting to expand the federal government's preemptive authority in the context of transmission siting. Congress sought through these provisions to empower the U.S. Department of Energy to designate "national interest transmission corridors" and to give FERC authority to preemptively approve the siting of transmission projects within such designated areas under certain prescribed circumstances.[25] Unfortunately, a pair of court decisions interpreting these provisions has cast doubt on their enforceability.[26] Despite years of effort to address the fragmented nature of transmission siting authority in the United States, the problem largely remains.

Recognizing that a more effective federal government solution to interstate transmission siting challenges for renewable energy is not likely to emerge in the near future, several U.S. states have begun working together toward developing alternative governance models. For example, 11 western U.S. states and two Canadian provinces have formed the Western Renewable Energy Zone Initiative, an advisory group that has collectively identified those areas in the western United States that it believes are most suitable for renewable energy development.[27] After several years of coordinated efforts, the group agreed upon a set of "hubs"—areas having concentrations of high-quality renewable resources and relatively low risks of environmental conflicts.[28] According to the group's materials, these hubs were identified "for purposes of evaluating interstate transmission lines in future phases of the initiative."[29]

Although efforts like the Western Renewable Energy Zone Initiative are laudable, they will never fully resolve interstate transmission siting difficulties because they do not consolidate the authority to fully approve multistate lines. Accordingly, some states have initiated discussions about going one step further and signing interstate compacts to create contractual one-stop siting approval systems.[30] At least one FERC commissioner has expressed support for this approach, which envisions using an interstate agreement to vest a single entity with the authority to approve all interstate transmission siting proposals confined within states that are signatories to the agreement.[31] Interstate compacts could eventually be an effective means of facilitating the interstate transmission build-out necessary to utilize the United States' vast supply of wind energy resources. Of course, the development of this innovative policy strategy is still in its embryonic stages and could face numerous political and other obstacles of its own.

Chickens, eggs, and "open season" siting schemes

In addition to the regulatory headaches just described, transmission planning in remote areas can also sometimes suffer from a classic chicken-and-egg problem: big wind and solar projects are only installed where there is adequate transmission capacity, but such capacity typically exists only where there are energy projects ready to use it. Developers of utility-scale renewable energy projects are hesitant to invest money in rural project sites that lack adequate grid access. At the same time, public utilities are hesitant to pursue new transmission projects in remote areas without high confidence that major wind or solar energy projects that would use them will be built there in the near future.[32] This unusual dynamic can also be an impediment to the build-out of transmission lines needed to fully support utility-scale renewable energy development in some countries.

Regulators in some nations of the world have used an "open season" siting scheme to address this chicken-and-egg problem and promote more optimal high-voltage transmission siting for renewable power. For example, the Bonneville Power Authority (BPA), which operates in the northwest United States, commenced such an open season program in 2008. Over an approximately three-month period, entities in that region were invited to submit requests for new high-voltage transmission capacity in whatever specific locations they desired.[33] All entities submitting such requests were required to commit in advance to actually purchase transmission capacity along any routes they had proposed if those lines were ultimately constructed.

The BPA's open season program enabled the agency to quickly gather a vast amount of high-quality information about which transmission projects would be most valuable to the region. The BPA used "cluster studies" to analyze the 153 requests it received and ultimately identified four major transmission project routes capable of being constructed at embedded cost rates.[34] Roughly 74 percent of the requested transmission capacity in this first open season was for proposed wind energy projects in the region.[35] Authorities in Mexico have used similar strategies to help better optimize transmission siting in that country.[36]

Of course, open season programs have their own limitations as well. Among other things, they are only effective to the extent that they have buy-in from all jurisdictions whose lands are involved. Still successful uses of this format by the BPA and Mexican authorities suggests that it can be a useful strategy for more efficiently planning long-distance transmission routes for renewable energy development in some remote areas.

Disputes over transmission infrastructure funding

Even if an interjurisdictional transmission line project has all necessary government approvals, it still may never materialize if stakeholders cannot agree on how to fund its construction. Extra-high-voltage transmission lines and supporting facilities are incredibly expensive. Estimations for one recent project in the United States put the cost of such lines at $1.7 million to $2.1 million per mile.[37] As other commentators have noted, disputes over who could cover these costs can be"just as much of a barrier" to the development of an interstate transmission line project as disputes over siting of the line itself.[38] It could cost up to $34 billion to expand transmission facilities enough to enable renewables to meet 33 percent of the annual electricity load in the western United States alone.[39] An estimated UK£110 billion of investment would be needed to adequately upgrade the U.K.'s electricity infrastructure.[40] Given the magnitude of these costs, it is no surprise that there is ongoing controversy about who should pick up this hefty tab.

A fundamental question frequently debated in disputes over transmission project funding is whether renewable energy *developers* should foot the bill or whether *utilities* should cover it. Laws in some jurisdictions have historically required energy developers to pay for the transmission infrastructure needed to serve their projects. For example, Australia has traditionally used a "developer pays" approach for transmission infrastructure, asking generators of electric power to finance the costs of connecting their generation facilities to the grid.[41] This strategy is akin to the relatively common practice of requiring developers of new residential subdivisions on the outskirts of a town to pay much of the expense of building new public roads to connect the new subdivision to existing city streets. Theoretically, such requirements are reasonable and equitable because real estate developers can pass along much of these costs to homebuyers.

Although this "developer pays" model may be an effective way of financing transmission facilities for conventional power plants near urban areas, the model is more prone to inefficiency in the context of renewable energy development. Utility-scale wind and solar energy projects tend to be in relatively remote locations, so the costs of transmission expansions necessary to connect them to the grid tend to be greater than for connecting conventional power plants. Using a "developer pays" model to fund transmission expansions for renewable energy projects is thus more likely to lead to "free rider" problems in which developers strategically wait to build their own projects in a remote region until some other developer has built a nearby line.[42] By waiting and being the second or third wind or solar energy project in a rural area, a patient renewable energy developer may be able to connect its own project into the newly-extended grid for a fraction of the cost it would have otherwise incurred. For obvious reasons, this sort of strategic waiting can inefficiently slow the pace of renewable energy development.

In contrast, some jurisdictions employ more of a "beneficiary pays" or "socialization" approach to fund new transmission lines. Under this approach, *utilities* generally finance most of the transmission improvements necessary to connect new renewable energy projects into the grid, even when such projects are not utility-owned. The utilities' customers— the beneficiaries—then ultimately pay for the infrastructure through their regular electricity bills.[43] Laws applicable in the U.S. state of Texas have long used a version of this sort of funding regime.[44]

Advocates of the "beneficiary pays" approach generally take the view that a well-functioning national electric grid is just as important as interstate freeway systems for automobiles or other interstate infrastructure projects that receive socialized funding and should thus be treated in the same way. Because this approach does not require renewable energy developers to directly fund transmission lines, even when their project is the first in a given area, it also evades the strategic waiting problems that can sometimes arise under "developer pays" regimes.

Unfortunately, "beneficiary pays" funding schemes can also be prone to controversy because it is not always clear who truly benefits from a new transmission line and should thus help to pay for it. Consider, for instance, a hypothetical transmission line project designed to increase interconnection between the windswept U.S. state of Montana and population centers in southern California. For purposes of determining who should pay for it under a "beneficiary pays" model, who would be the "beneficiaries" of this new line? Californians? Montanans? Someone else?

One cannot resolve the question of who "benefits" from a new transmission line by merely following the flow of the electric current itself. Given the complexity and interwoven nature of the United States' "Western Interconnection" grid system, it is impossible to route specific electrons generated on wind farms in Montana to particular California homes and businesses several hundred miles away. The reality is that much of the electricity generated at the wind farms would likely flow to electricity users in Montana or neighboring states rather than to California, but no one would argue that Montanans should solely pay for the transmission system on that ground. In fact, the additional transmission capacity provided by the new line could ease grid operations in multiple U.S. states in addition to Montana and California, thereby benefiting electricity users in those other states as well. Should these other states have to help finance the line, even though they are not directly involved in its planning or construction? A traditional "beneficiary pays" model does not squarely answer these sorts of inquiries.

"Beneficiary pays" funding models for interjurisdictional transmission can also be vulnerable to their own strategic behavior problems. For example, suppose that a law applying the model required all users of a large interconnected grid to collectively share the costs of new transmission lines regardless of their locations. Under such a law, some jurisdictions might be incentivized to advocate for new transmission development within their boundaries solely to get local economic boosts at the expense of others—the transmission equivalent of the infamous "bridge to nowhere."[45] To quote an Australian official on this risk of socialization-based cost allocation schemes:

> [I]f transmission infrastructure building costs were to be recovered …
> from all end users, this could distort the operation of the market by
> encouraging generation connections in areas that are inefficient or more
> expensive than elsewhere.[46]

In an attempt to address the incessant controversy surrounding cost allocation schemes for interstate transmission projects in the United States, FERC issued Order 1000 in 2011.[47] This administrative order sets forth several vague "cost allocation principles" and essentially requires that all new major transmission projects comply with them.[48] For example, the Order requires that cost allocations associated with any large transmission project be "roughly commensurate" with benefits received.[49] Under the Order, FERC purportedly has authority to step in and determine a cost allocation method in situations when regions are unable to agree to such a method on their own.[50]

Although FERC Order 1000 has the potential to reduce conflicts over cost allocation arrangements for interstate transmission projects, its ultimate impact remains to be seen. Order 1000's ambiguous cost allocation principles create significant uncertainty about how the Order is to be applied and enforced among the various public utilities and regional system operators that collectively oversee electric grid operations in the United States. To make matters worse, some of these regional operators view Order 1000 as an encroachment on their regulatory powers and have thus sought to use the judicial system to limit the Order's effects. As of late 2013, at least one federal appeals court had largely upheld Order 1000's cost allocation requirements.[51] However, a claim challenging the Order in a different federal court was casting doubts about the ultimate fate of this latest attempt by FERC to ease cost allocation negotiations for large scale transmission projects.[52] It remains to be seen whether the Order will ultimately prove successful at reducing controversies in the United States over the funding of new interstate transmission lines.

On a more optimistic note, increasing amounts of private investment in the energy sector offer some hope that disputes over the funding of high-voltage transmission infrastructure could decrease in years to come. If these sorts of private capital investments grow to become a more prominent funding source, they could alleviate at least some of the present struggles over transmission funding. For example, the Internet behemoth Google has invested heavily in the $5 billion Atlantic Wind Connection project—a planned high-voltage "backbone" line aimed at facilitating up to 6 GW of offshore wind energy development along the eastern seaboard of the United States.[53] Google executives have suggested that the company intends to invest in more "transformative" transmission projects in the future.[54] If this trend were to continue and similar companies were to join it, many of the cost allocation challenges described above would become less relevant over time, facilitating a faster pace of renewable energy development.

Conflicts over renewable energy's impacts on grid operations

A second category of grid-related conflicts that can hinder renewable energy development relates not to the *transmission* of electricity across long distances but to its *distribution* via lower-voltage lines to end-users. Like the transmission-related conflicts just described, many distribution-related disputes arise because existing grid facilities and operations in many regions were not originally designed to manage the unique characteristics of renewable power. In particular, existing grids are seldom equipped to effectively handle the intermittent nature of many renewable energy resources.

The availability of wind and solar energy resources not only varies by location—it also varies by time. The wind does not always blow, and the sun does not always shine. For the most part, only Mother Nature can control fluctuations in the quality of wind and solar resources at a given location. These fluctuations can be unpredictable and can sometimes occur in a dramatic fashion and on relatively short notice.

The intermittency and variability of wind and solar energy resources can complicate efforts to incorporate more of these resources into a utility's energy mix. Stable, malleable power sources have long been instrumental in enabling utilities to provide reliable electricity service to consumers without blackouts or other interruptions. As one writer describes it:

On an electricity grid, supply and demand must be balanced continuously to maintain a variety of physical network criteria—like frequency, voltage, and capacity constraints—within narrow bounds. Electricity is the ultimate "just in time" manufacturing process, where supply must be produced to meet demand in real time.[55]

Historically, grid operators have had a difficult enough time using conventional coal—nuclear—or natural gas-fired power plants to balance electricity supply and demand on electric grids. The aggregate quantity of power demanded across a grid system varies from moment to moment as electricity consumers respond to such factors as the outside temperature or the time of day. Grid operators have little or no control over these minute-by-minute changes in electricity demand, but fortunately they can exert some control over the quantity of electricity that conventional power plants supply to the grid. By ramping electricity production up and down at these plants in response to real-time variations in electricity demand, utilities and grid operators are typically able to ensure that no one goes without power and that no excessive power overloads damage the grid.[56]

Allowing wind and solar energy systems to feed electricity onto an electric grid makes grid operators' perpetual balancing act at least twice as difficult to perform. Adding intermittent renewable energy sources into the power mix can force grid operators to simultaneously respond to fluctuations in both the demand *and* the supply of electricity on the grid. Utilities confronted with this arduous task have sometimes responded by cutting back their electricity purchases from renewable energy projects that they perceive to be the source of their problems. Whenever this occurs, disputes between renewable energy generators and grid operators are bound to follow.

Publicized clashes between wind energy generators and grid operators in the northeastern United States highlight the tension that can build between renewable energy generators and grid operators as renewable energy systems feed ever more power onto the grid. To prevent the variability of wind-generated electricity from damaging its transmission infrastructure, ISO New England has been known to block wind-generating power from entering its grid systems during periods when electricity supply far exceeds demand. These "curtailment" practices have cost wind energy generators in the region of millions of dollars, infuriating the owners of multiple wind projects.[57] As an angry wind energy executive from the U.S. state of Vermont put it:

We have a grid system that's not smart—it's kind of dumb, it's a 100-year-old system—and they run it like fossils and nukes are the only things that matter and the rest of us, they can fiddle with.[58]

Easing grid management through smarter grids

The most promising way to reduce grid management difficulties in an era of wind and solar energy is to make the grid itself "smarter" and better equipped to quickly adapt to fluctuations in electricity supply and demand. In the case involving the Vermont wind farms described above, the parties ultimately determined that installing a new piece of equipment could largely ameliorate their problem.[59] Innovative enhancements to grids aimed at dealing with these sorts of disruptions are commonly referred to as "smart grid" technologies.

A fully updated smart grid can provide almost instantaneous information to grid operators regarding the amounts of electricity being supplied and consumed at any given time and can respond quickly and effectively to changes. The following definition by the Electric Power Research Institute (EPRI) more completely describes the concept of smart grid technology:

> The term "Smart Grid" refers to the modernization of the electricity delivery system so that it monitors, protects, and automatically optimizes the operation of its interconnected elements—from the central and distributed generator through the high-voltage transmission network and the distribution system, to industrial users and building automation systems, to energy storage installations, and to end-use consumers, and their thermostats, electric vehicles, appliances, and other household devices.[60]

A smart grid that fully fits the definition above is presently little more than an idealized, "pie-in-the-sky" fantasy in most of the world. The EPRI estimates that fully building out such a grid in the United States alone would require a net investment of between $338 billion and $476 billion.[61] The U.S. seems unlikely to invest anything near that sum of money in grid infrastructure improvements anytime soon.

Still, even modest upgrades focused on specific smart grid concepts could do much to mitigate the intermittency and variability challenges raised by the addition of wind and solar energy systems to the grid. The U.S. could upgrade only its high-voltage transmission infrastructure with "smart" devices for roughly $60 billion, and researchers in this area believe such upgrades would be the "most cost-effective category of smart grid investments."[62] In the U.S. or in any country, these sorts of upgrades can dramatically improve system operators' abilities to balance electricity supply and demand as more and more renewable energy systems begin delivering power onto grid systems. Such benefits make smart grid innovations an essential element of any plan to prevent grid disruption concerns from slowing the long-term growth of wind and solar energy.

Opposition from utilities that view renewables as a competitive threat

In addition to complicating grid operations for utilities, the renewable energy sector is increasingly creating fiscal challenges for utilities as well. As wind and solar energy have grown in the past decade, some utilities and traditional energy companies that formerly seemed to favor renewable energy development have become vocal opponents of it. Once little more than a fledgling side industry good for bolstering utilities' public relations, renewable energy has emerged as a formidable threat to utilities' profitability and long-term existence in some markets. The growing clashes between utilities and renewable energy interests are manifestations of the friction inevitably involved in transforming an entrenched but outmoded electricity industry into something far more financially and ecologically sustainable. Because these clashes have more to do with economics than logistics, they do not easily lend themselves to technological solutions. Consequently, they can be among the most perplexing and difficult conflicts to resolve.

Lobbying against renewables

For years, major stakeholders in the electricity industry have openly lobbied against renewable energy in hopes of protecting their own economic interests. In some instances, these protectionist motives have been known to drive opposition to wind energy in the United States. Once a wind energy system is up and running, it can produce electricity for decades at almost zero marginal cost. During seasons and hours of the day when energy demand is low, the growing abundance of wind-generated energy in some electricity markets is driving down spot prices and cutting deeply into the conventional power generators' pocketbooks. Hoping to avoid these financial losses, a growing number of longtime players in the electric power sector who once claimed to support renewable energy are quickly becoming its political foes.

   Exelon Corp., an energy company in the United States, is one example of a utility that has morphed from an avid supporter of the renewable energy sector to an active opponent of it. Exelon operates as a regulated utility in multiple U.S. states and owns several nuclear energy generating facilities. The company was once a member of the American Wind Energy Association (AWEA) and had even touted its own wind energy holdings and commitment to sustainability in its marketing materials.[63] However, as more and more wind farms began sprouting up across the mid-western United States, spot prices for electricity in the region dropped during certain periods and cut into the company's profits.

   Frustrated by the economic impacts of wind energy development on the regional electricity market, Exelon became a vocal advocate and lobbyist against renewal of a federal tax credit program that was critical to the continued rapid growth of the U.S. wind energy industry.[64] In late 2012, AWEA responded to Exelon's sudden pivot on wind energy issues by kicking Exelon out of the group.[65] Although Congress ultimately extended the tax credits for one year, Exelon continued its fight. As of mid-2013, the company was lobbying to block yet another extension of the popular tax credit program.[66] As wind and solar energy grow ever more cost-competitive with conventional electricity, these industries are often encountering a growing number of political enemies within the energy industry.

Anticompetitive tactics against renewables

A handful of utilities have gone beyond lobbying and actually leveraged their authority and control over grid operations to discourage the rise of renewable energy within their territories. In efforts to preserve their long-held dominance of electricity markets, utilities in some countries have discriminated against renewable energy developers in ways that make it difficult or impossible for them to connect their projects into the grid. For example, transmission companies in the Philippines have been known to charge excessive rates for transmitting wind- or solar-generated power.[67] Grid operators are rumored to use similar sorts of tactics in Thailand as well.[68]

Hoping to prevent utilities' anticompetitive behavior from hampering the European transition to renewable energy, policymakers in Europe have enacted provisions in the past decade that prohibit discrimination against renewable energy providers in their requests for access to the electric grid. In 2009, the European Union adopted a directive containing provisions that address this issue. Part 63 of the preamble of Article 16 of European Directive 2009/28/CE reads:

> Electricity producers who want to exploit the potential of energy from renewable sources in the peripheral regions of the Community ... should, whenever feasible, benefit from reasonable connection costs in order to ensure that they are not unfairly disadvantaged in comparison with producers situated in more central, more industrialized and more densely populated areas.[69]

The EU directive itself contains provisions aimed at implementing this principle. Among other things, the directive expressly requires that the "charging of transmission and distribution tariffs [sic] not discriminate against electricity from renewable energy sources"[70] and that such sources receive "either priority access or guaranteed access to the grid system."[71]

Policies protecting grid access privileges for renewables are even stronger in Germany, where laws require "grid system operators" to "immediately and as a priority connect installations generating electricity from renewable energy sources."[72] In countries where grid access obstacles are slowing the growth of renewable energy, these sorts of provisions can be a potentially effective and relatively straightforward way of addressing such problems.

Net metering and the grid funding problem

The most perplexing controversies brewing between utilities and the renewable energy industry relate to the impact of net metering programs on the long-term financial stability of utility companies. Although net metering programs can do much to incentivize distributed generation and energy efficiency, they also have the potential to ultimately wreak havoc on utility budgets. Fearful of these impacts, a growing number of utilities are challenging net metering programs in ways that could severely hamper distributed renewable energy development.

Net metering programs exist in several nations throughout the world and were active in more than 40 U.S. states as of late 2013.[73] These programs typically allow owners of residential and commercial-scale renewable energy systems to receive credit for any excess electricity their systems produce that gets fed onto the grid. Owners can then use these credits to directly offset their own electricity bills.[74] Landowners with rooftop solar energy systems are the most common participants in net metering programs. Depending on their electricity usage and on the size of their rooftop solar array, solar users can potentially reduce their electricity bills to nearly zero through these programs—a compelling incentive to go solar.

The problem is that, as the number of solar users within a given utility's territory grows, the utility sells less and less conventional power to rate-paying customers and thus brings in less and less revenue. In the United States and in other jurisdictions with comparable utility rate-setting structures, this revenue drop prompts utilities to seek approval from state regulators to charge ratepayers higher per-kilowatt-hour prices for power. These petitions for rate increases are generally justifiable: a utility that is selling less and less electricity can maintain its present rate of return on its capital investments only by charging more and more for each kilowatt-hour of power it sells.

Unfortunately for utilities, electricity rate hikes aimed at accounting for a growing number of solar energy users only further strengthen the competitive position of the solar energy industry. For obvious reasons, buying or leasing solar panels and participating in net metering becomes more financially rewarding for ratepayers as conventional electricity rates rise. Rate increases thus tend to incentivize even more solar panel installations that further reduce utilities' volume of electricity sales. The only way for utilities to maintain revenues in response to such sales reductions is through additional rate increases that serve only to make solar energy even more attractive to ratepayers. If this cycle continues unabated, it can send electricity rates spiraling ever upward and eventually become an existential threat to utilities that had historically been pinnacles of financial stability.

Spain's experience with net metering and related programs aimed at promoting renewable energy exemplifies the financial risks that such programs can pose for conventional utilities. Spain was long lauded for being near the forefront among the nations of the world in promoting renewable energy development through net metering, feed-in tariffs and other laws that subsidized renewables. The nation's laws were wildly successful in incentivizing a rapid and dramatic increase in solar energy installations within that country. As of April 2013, 54 percent of all electricity generated in Spain came from renewables.[75]

Unfortunately, this swift transition to sustainable energy proved to be anything but sustainable. By May of 2013, Spain's "tariff deficit" under its feed-in tariff program was roughly $34 billion and growing.[76] The quantity of conventional electricity being sold to ratepayers in Spain was nowhere near enough to cover the costs associated with financing and maintaining its electric grid. By late 2013, the nation was mulling a tax on solar energy users as a way of restoring financial solvency in its energy sector.[77] The proposed tax would potentially apply retroactively even to individuals who had purchased solar energy systems several years earlier. It would also make owning solar energy systems less financially rewarding than reliance on conventional electricity delivered via the nation's electric grid.[78] If adopted, such policy changes would instantly convert Spain from one of the most solar energy-friendly countries in the developed world to one of the least supportive of solar energy technologies.

Because of their potential to threaten the financial stability of utilities, net metering programs are beginning to stir controversy in the United States as well. Utilities have already unsuccessfully challenged net metering policies in the U.S. states of Idaho and Louisiana.[79] By mid-2013, the debate had moved to the states of Arizona and California—the top two states for total installed solar energy generating capacity.[80] Government officials in California ultimately responded to this debate in October of 2013 by authorizing utilities to charge solar users fees of up to $10 per month to ensure that such users paid a fair share of the costs of maintaining the electric grid.[81]

In Arizona, a large utility company sought in late 2013 to weaken the state's net metering programs and pursued state regulatory approval to impose fees of $100 or more on solar users.[82] The utility argued that its proposed changes were reasonable policy steps designed to ensure that additional grid maintenance costs didn't fall on non-solar electricity customers as solar energy installations increased over time. Not surprisingly, solar industry advocates disagreed and characterized the changes as a "tax on the solar energy industry" that would "effectively end the roof-top solar market" in Arizona.[83] State regulators ultimately rejected the utility's specific proposals but did vote to allow a monthly fee of $0.70 per kilowatt of installed solar generating capacity on the electricity bills of customers who installed solar energy systems after December 31, 2013.[84] Importantly, all customers who installed solar energy systems on their properties after that date would also be required to sign a document acknowledging that such fees could increase at any point in the future.[85] Because this particular provision creates tremendous uncertainty for those contemplating solar energy in Arizona, it is likely to have a chilling effect on distributed solar energy development in that state.

As this book was heading into publication, several more states in the United States appeared poised to consider proposals to modify their net metering programs or to impose fees or taxes on solar energy users in the near future.[86] Given the inevitable tension between net metering and traditional utility funding mechanisms, political conflicts between utilities and the renewable energy sector seem unlikely to disappear anytime soon.

Renewables and the future of the electric grid

In spite of all of the challenges described in this chapter, there is reason to believe that human ingenuity and determination will ultimately enable society to overcome present conflicts between grids and renewable energy. For example, 78 percent of the electricity consumed annually in the United States is used in the 28 states that border the coasts.[87] Relatively little new transmission infrastructure would be needed to deliver electricity from offshore wind energy projects off those states' coastlines to urban centers. A greater focus on offshore wind energy in the United States could thus reduce the need for additional transmission capacity to deliver power from wind farms in remote interior regions to coastal cities. Some private investors have even shown interest in helping to fund the development of offshore transmission facilities to help support this strategy. The Atlantic Wind Connection project—a proposed transmission "backbone" off of the coast of the eastern United States—counts the Internet giant Google as one of its major financial backers.[88]

Further advancements in energy storage technologies, which are discussed in more detail in Chapter 9, could also eventually do much to ease the current tension between renewables and utilities. Innovations allowing owners of solar or wind energy systems to cheaply store excess power generated during sunny or windy times could help mitigate grid load balancing problems resulting from the intermittent nature of those resources. Affordable energy storage could also make having a connection to the electric grid less crucial or even unnecessary for many households and businesses.

Until these sorts of technological innovations become available, *policy* innovations can do much in their own right to prevent grid-related constraints from unduly hampering the renewable energy movement. Policymakers can greatly assist the global sustainable energy transition by developing policies that are well tailored to the unique transmission and distribution challenges associated with renewables. More policies akin to the open season programs and interstate compact arrangements described above are needed to facilitate the efficient build-out of transmission capacity in remote areas that will be necessary to support full development of the world's wind and solar development. Within urban areas, thoughtful revisions to electricity rate setting policies will also grow ever more critical to efficiently governing conflicts between the rooftop solar energy industry and electric utilities.

Renewable energy technologies are ever improving, so it is not inconceivable that a day will eventually arrive when the electric grid as we know it is no longer needed at all. Like the pony express of nineteenth-century America or the antiquated telegraph infrastructure that succeeded it, electricity grids seem destined to eventually be supplanted by something better. If innovations someday eliminate the need to interconnect energy users via vast, complex webs of lines and substations, many of the struggles described in this chapter will fall by the wayside, greatly simplifying sustainable energy development. Until that vision is achieved, however, engineers, policymakers and industry players each have much to contribute in the global effort to address grid-related challenges so that the global transition toward more sustainable energy sources can continue unabated throughout this century and beyond.

Notes

1 David B. Spence, *Regulation, Climate Change, and the Electric Grid*, 3 SAN DIEGO J. CLIMATE & ENERGY L. 267, 271 (2012).

2 Id.

3 Other researchers have similarly emphasized challenges raised by the immobile nature of wind resources. *See, e.g.*, Sandeep Vaheesan, *Preempting Parochialism and Protectionism in Power*, 49 HARV. J. ON LEGIS. 87, 97 (2012) (noting that, "[u]nlike coal, natural gas, and other fossil fuels, which can be moved by pipeline, rail or ship, renewable energy resources are location-specific and cannot be transported to points closer to load centers and used to generate power there").

4 *See* American Wind Energy Association, *Top 20 States with Wind Energy Potential* (2008), available online at www.casperlogisticshub.com/downloads/Top_20_States.pdf (last visited Sept. 3, 2013).

5 U.S. Census Bureau, *Statistical Abstract of the United States: 2012* at 19 (2012), available at www.census.gov/compendia/statab/2012/tables/12s0014.pdf (last visited Sept. 5, 2013).

6 *See, e.g.*, Potomac Economics, *2012 State of the Market Report for the Midwest ISO* at 49 (2013) (noting that "continued increases in wind output in 2012 resulted in more congestion on constraints carrying power out of the West").

7 Anne Kallies, *The Impact of Electricity Market Design on Access to the Grid and Transmission Planning for Renewable Energy in Australia: Can Overseas Examples Provide Guidance?*, 2 RENEWABLE ENERGY L. & POL'Y REV. 147, 158 (2011).

8 *See* Dermot Duncan & Benjamin K. Sovacool, *The Barriers to the Successful Development of Commercial Grid Connected Renewable Electricity Projects in Australia, Southeast Asia, the United Kingdom and the United States of America*, 2 RENEWABLE ENERGY L. & POL'Y REV. 283, 295 (2011).

9 *See* Vaheesan, *supra* note 3 at 113–14.

10 *See generally* Stale Navrud et al., *Valuing the Social Benefits of Avoiding Landscape Degradation from Overhead Power Transmission Lines: Do Underground Cables Pass the Benefit-Cost Test?*, 33 LANDSCAPE RES. 281 (2008).

11 *See* id. at 112–13.

12 For an excellent discussion of microgrids and their potential as a way of avoiding energy sprawl, *see generally* Sara C. Bronin, *Curbing Energy Sprawl with Microgrids*, 43 CONN. L. REV. 547 (2010).

13 *See* Alan Goodrich, et al., *Residential, Commercial, and Utility-Scale Photovoltaic (PV) System Prices in the United States: Current Drivers and Cost-Reduction Opportunities* at 20, National Renewable Energy Laboratory Technical Report NREL/TP-6A20-53347 (Feb. 2012) (showing utility-scale PV system costs at between $3.80 and $4.40 per watt, compared to $4.59 per watt for commercial-scale PV systems).

14 *See* Andrew Mills, et al., *Exploration of Resource and Transmission Expansion Decisions in the Western Renewable Energy Zone Initiative, Berkeley National Laboratory* at 13, LBNL-3977E (2010), available at http://eetd.lbl.gov/EA/EMP/reports/lbnl-3077e.pdf (last visited Aug. 22, 2013).

15 *See* Deborah Behles, *An Integrated Green Urban Electrical Grid*, 36 WM. & MARY ENVTL. L. & POL'Y REV. 671, 677–78 (2012).

16 *See* The Honorable John R. Norris & Jeffery S. Dennis, *Electric Transmission Infrastructure: A Key Piece of the Energy Puzzle* at 5, NAT. RESOURCES & ENV'T (Spring 2011)

17 *See* Norris & Dennis, *supra* note 16 at 6.

18 Ashley C. Brown & Jim Rossi, *Siting Transmission Lines in a Changed Milieu: Evolving Notions of the "Public Interest" in Balancing State and Regional Considerations*, 81 U. COLO. L. REV. 705, 717 (2010).

19 *See* Vaheesan, *supra* note 3 at 116.

20 *See* Brown & Ross, *supra* note 18 at 724–25.

21 Paul Davenport, *Arizona Regulators Skeptical of New Electrical Line to California*, U-T SAN DIEGO (2007), available at http://legacy.utsandiego.com/news/business/20070530-1352-wst-sharingpower.html (last visited Sept. 19, 2013).

22 16 U.S.C. § 824(a).

23 Id.

24 *See* Jim Malewitz, *Wind, Solar Could Benefit from Kansas Transmission Compact*, MIDWEST ENERGY NEWS (July 29, 2013), available at www.midwestenergynews.com/2013/07/29/wind-solar-could-benefit-from-kansas-transmission-compact/ (last visited Sept. 19, 2013).

25 16 U.S.C. § 824p.

26 *See Piedmont Environmental Council v. FERC*, 558 F.3d 304 (4th Cir. 2009) (holding that FERC's authority to site a transmission project under 16 U.S.C. § 824p does not apply if the state rejects the project permit application within one year of its filing); *California Wilderness Coalition v. Dep't of Energy*, 631 F.3d 1072 (9th Cir. 2011) (invalidating a Department of Energy designation of a "national interest transmission corridor").

27 *See* Hannah Wiseman, *Expanding Regional Renewable Governance*, 35 HARV. ENVTL. L. REV. 477, 516 (2011).

28 To download the Western Renewable Energy Zones Phase I Map and read more about the initiative, visit the Western Governors Association's website at www.westgov.org/component/content/article/102-initiatives/219-wrez (last visited Sept. 19, 2013).

29 Id.
30 *See* Malewitz, *supra* note 24.
31 *See* id.
32 At least two other scholars have highlighted this "chicken-and-egg" problem in the context of interregional transmission siting. *See, e.g.*, Jennifer E. Gardner & Ronald L. Lehr, *Wind Energy in the West: Transmission, Operations, and Market Reforms*, 26 WTR NAT. RESOURCES & ENV'T 13 (Winter 2012); Glen Wright, *Facilitating Efficient Augmentation of Transmission Networks to Connect Renewable Energy Generation: The Australian Experience*, 44 ENERGY POLICY 79, 80 (2012).
33 *See* Allison Schumacher, et al., *Moving Beyond Paralysis: How States and Regions Are Creating Innovative Transmission Policies for Renewable Energy Projects*, 22 ELECTRICITY J. 27 (2009).
34 *See* id.
35 *See* id.
36 *See generally* Francisco J. Barnes, *An Open Season Scheme to Develop Transmission Interconnection Investments for Large Wind Farms in Mexico*, Comision Reguladora de Energia (2009), available at http://siteresources. worldbank.org/INTENERGY/Resources/335544-1232567547944/5755469-1239633250635/Francisco_J_Barnes.pdf (last visited Feb. 24, 2014).
37 *See* Miriam Sowinski, *Practical, Legal, and Economic Barriers to Optimization in Energy Transmission and Distribution*, 26 J. LAND USE & ENVTL. L. 503, 521 (2011) (describing a cost estimate of a project by American Electric Power).
38 Brown & Rossi, *supra* note 18 at 758.
39 *See* Mills, *supra* note 14 at v.
40 *See* Duncan & Sovacool, *supra* note 8 at 296.
41 *See* Kallies, *supra* note 7 at 152.
42 *See* Wright, *supra* note 32 at 80–81.
43 *See* Norris & Dennis, *supra* note 16 at 6.
44 *See* Gardner & Lehr, *supra* note 32 at 14 (Winter 2012) (citing Matthew H. Brown & Richard P. Sedano, *Electricity Transmission: A Primer* 22 (2004)).
45 For a description of the proposed "bridge to nowhere" and its ultimate demise, *see generally* Carl Hulse, *Two "Bridges to Nowhere" Tumble Down in Congress*, NEW YORK TIMES at A19 (Nov. 17, 2005).
46 *See* Kallies, *supra* note 7 at 153–54.
47 *Transmission Planning and Cost Allocation by Transmission Owning and Operating Public Utilities*, 136 FERC ¶ 61051 (2011).
48 Id.
49 Id. at 440.
50 Federal Energy Regulatory Commission, *Briefing on Order 1000*, Slide 14 (2011), available at www.ferc.gov/media/news-releases/2011/2011-3/07-21-11 E-6-presentation.pdf (last visited Sept. 24, 2013).
51 *See generally Illinois Commerce Commission v. FERC*, 2013 WL 2451766 (2013).
52 Sonal Patel, *Challenges to Order 1000 Filed in Federal Court as President Acts on Grid Modernization*, POWER (June 13, 2013), available at www.powermag. com/correctedchallenges-to-order-1000-filed-in-federal-court-as-president-acts-on-grid-modernization/ (last visited Sept. 24, 2013).
53 Ros Davidson, *Grid Must Catch Up with Renewables Growth*, WINDPOWER MONTHLY (Sept. 1, 2013), available at www.windpowermonthly.com/ article/1209701/grid-catch-renewables-growth (last visited Sept. 24, 2013).
54 Id.

55 Paul L. Joskow, *Creating a Smarter U.S. Electricity Grid*, 26 J. ECON. PERSPECTIVES 29, 31 (2012).
56 *See* Spence, *supra* note 1 at 274.
57 *See* Diane Cardwell, *Intermittent Nature of Green Power is Challenge for Utilities*, NEW YORK TIMES B1 (Aug. 15, 2013).
58 Id.
59 *See* id.
60 Electric Power Research Institute, *Estimating the Costs and Benefits of the Smart Grid: A Preliminary Estimate of the Investment Requirements and Resultant Benefits of a Fully Functioning Smart Grid* 1-1, Tech. Rep. No. 10225519 (2011), available at http://ipu.msu.edu/programs/MIGrid2011/presentations/pdfs/Reference%20Material%20-%20Estimating%20the%20Costs%20and%20Benefits%20of%20the%20Smart%20Grid.pdf (last visited Oct. 4, 2013).
61 *See* id. at 1–4.
62 *See* Joskow, *supra* note 55 at 36.
63 *See* Julie Wernau, *Exelon Pushes to Scrap Wind Subsidy*, CHICAGO TRIBUNE (Aug. 3, 2012).
64 *See* id.
65 *See* Julie Wernau, *Wind Energy Group Gives Exelon the Boot*, CHICAGO TRIBUNE (Sept. 10, 2012).
66 Jim Snyder & Julie Johnsson, *Exelon Falls from Green Favor of Chief Fights Wind Aid*, BLOOMBERG.COM (Aug. 1, 2013), available at www.bloomberg.com/news/2013-04-01/exelon-falls-from-green-favor-as-chief-fights-wind-aid.html (last visited Oct. 3, 2013).
67 *See* Duncan & Sovacool, *supra* note 8 at 292.
68 *See* id.
69 *Directive 2009/28/EC of April 2009 on the Promotion of the Use of Energy from Renewable Sources and Amending and Subsequently Repealing Directives 2001/77/EC and 2003/30/EC, OJ L 140/19* (2009) (cited in Kallies, *supra* note 7 at 156) (available at http://eur-lex.europa.eu/LexUriServ/LexUriServ.do?uri=Oj:L:2009:140:0016:0062:en:PDF (last visited Sept. 24, 2013).
70 Id. at Art. 16, § 7.
71 Id. at Art. 16, § 2(b).
72 *Energiewirtschaftsgesetz 2005*, § 17–19 (cited in Kallies, *supra* note 7 at 157), available in translated English online at https://financere.nrel.gov/finance/node/2740 (last visited Sept. 24, 2013).
73 *See* U.S. Dept. of Energy, *Net Metering*, Database of State Initiatives for Renewables and Efficiency, available at www.dsireusa.org/solar/solarpolicyguide/?id=17 (last visited Oct. 4, 2013).
74 *See* id.
75 *See* William Pentland, *No End in Sight for Spain's Escalating Solar Crisis*, FORBES (Aug. 16, 2013), available at www.forbes.com/sites/williampentland/2013/08/16/no-end-in-sight-for-spains-escalating-solar-crisis/ (last visited Oct. 8, 2013).
76 *See* id.
77 *See* Mira Galanova, *Spain's Sunshine Toll: Row Over Proposed Solar Tax*, BBC NEWS (Oct. 6, 2013), available at www.bbc.co.uk/news/business-24272061 (last visited Oct. 8, 2013).
78 *See* id.
79 *See* Mark Jaffe, *Rooftop Solar Net Metering Is Being Fought Across the U.S.*, DENVER POST (Sept. 1, 2013), available at www.denverpost.com/business/ci_23986631/rooftop-solar-net-metering-is-being-fought-across (last visited Oct. 8, 2013).

80 *See* GTM Energy and Solar Energy Industries Association, *U.S. Solar Market Insight* 5 (2012), available at www.seia.org/sites/default/files/resources/ZDgLD2dxPGYIR-2012-ES.pdf (last visited Oct. 8, 2013).

81 *See* Chris Clarke, *California Governor Signs Controversial Solar Bill, Among Others*, KCET.org (Oct. 7, 2013), available at www.kcet.org/news/rewire/government/governor-signs-controversial-solar-bill.html (last visited Oct. 8, 2013).

82 *See* Scott Shugarts, *AZ Solar Customers Face New Proposed Fees*, EnergyBiz.com (Sept. 5, 2013), available at www.energybiz.com/article/13/09/az-solar-customers-face-new-proposed-fees (last visited Oct. 8, 2013).

83 Id.

84 *See* Ryan Randazzo, *Commission Votes to Raise APS Solar Customers' Bills*, ARIZONA REPUBLIC (Nov. 14, 2013), available at www.azcentral.com/business/arizonaeconomy/articles/20131114aps-solar-customer-bills-higher.html (last visited Dec. 19, 2013).

85 *See* id.

86 *See* Herman K. Trabish, *Rooftop Solar Scores a Crucial Win Against Arizona's Dominant Utility*, GreenTechMedia.com (Oct. 3, 2013), available at www.greentechmedia.com/articles/read/Crucial-Win-For-Rooftop-Solar-Over-Arizonas-Dominant-Utility (last visited Dec. 19, 2013).

87 *See* Erica Schroeder, *Turning Offshore Wind On*, 98 CAL. L. REV. 1631, 1640 (2010).

88 *See* Dave Levitan, *Will Offshore Wind Finally Take Off on U.S. East Coast?*, YALE ENVIRONMENT 360 (Sept. 23, 2013), available at http://e360.yale.edu/feature/will_offshore_wind_finally_take_off_on_us_east_coast/2693/ (last visited Oct. 16, 2013).

8 Overcoming obstacles in wind energy development

A case study

Contrary to what the previous several chapters might suggest, the message of this book is not that renewable energy development is too laden with obstacles and challenges to be worth the effort. True, some wind turbines pose hazards to birds or create irritating flicker effects. Rooftop solar energy systems can stick out like a sore thumb in some historic neighborhoods. However, despite these sorts of issues, valuable wind and solar energy development is proliferating at an astounding rate throughout the world. Every day, developers are finding ways to navigate through the countless conflicts and difficulties highlighted in this book, actively guiding renewable energy projects to completion. There has arguably never been a more promising time for energy technology over the course of human history, and the prospects for future growth have never been brighter.

Of course, to state that renewable energy development is often doable and worthwhile is not to imply that it is easy. Anyone who has had heavy involvement in a large wind or solar energy project can attest to the high degree of determination that developers and their teams must have to succeed. For the largest projects, the work associated with identifying a favorable project site, securing the requisite property interests and financing, gaining all necessary approvals, and ultimately completing construction can take a decade or longer. Experienced renewable energy developers seem to embrace the heavy workload and incessant challenges inherent in their industry. To them, all of the effort is worth it in the end, when they make their first visit to a completed project that has at last begun generating clean, renewable electricity from the wind or sun.

The developers who are most successful at shepherding large wind and solar energy projects from start to finish tend to view them as more than mere financial investments for economic gain. They recognize that, although wind farms and solar energy projects generally benefit the planet as a whole, they can also greatly impact the lives of real people living nearby. Accordingly, the savviest of developers understand the value of listening sincerely to stakeholder concerns, encouraging open dialogue with adversaries, and generously giving back to host communities.

This chapter offers an in-depth account of the long journey that one renewable energy developer took to develop a large wind energy project: the 500 MW Windy Point/Windy Flats project in the United States. There is nothing particularly unique or remarkable about this wind farm or how it was developed; a similar sort of story could probably be told about dozens of other successful projects. However, by delving deeply into the events surrounding the development of this single wind farm—the personalities and places—this chapter seeks to highlight the complex and important human aspect that is inevitably present in most major renewable energy projects.

A windy, welcoming place

Klickitat County, Washington, shown in Figure 8.1 below, is a rural county nestled on the eastern slopes of the Cascade Mountains in the northwest United States. The county's western edge is covered with rugged, forested terrain. A traveler heading east across Klickitat County is bound to be struck by the rapidly changing landscape, with pine-topped mountains gradually giving way to rolling foothills and then to grassy prairies and plateaus. This quiet pocket of south-central Washington has remained largely the same for more than a century, retaining a charming, small-town character.

The Columbia River Gorge—a majestic river canyon carved deep into the Cascades—forms Klickitat County's southern border. Each year, hundreds of thousands of wild Chinook salmon migrate up the Columbia River

Figure 8.1 Klickitat County, Washington, United States

from the Pacific Ocean to spawn. Coincidentally, steady wind currents often follow the same general path as the salmon, coursing more than 100 miles up the Gorge due to major differences between ocean and inland air temperatures. In Klickitat County, these "Chinook winds" tend to be the strongest on ridge tops within roughly three miles of the Gorge's north rim. Without question, they comprise some of the richest wind resources in the northwest United States.[1]

Many of the approximately 20,000 residents of Klickitat County have long been accustomed to the frequent windiness in the area. For most of the county's history, these winds had never amounted to much more than an annoyance—just another element of rough-and-tough ranching life. A cattle rancher in the county once recounted how he routinely battled strong headwinds as a child while pedaling his bicycle to school.[2] He and his friends eventually learned that these gusty conditions were more manageable if they zigzagged back and forth across the road on their bikes as they traveled. They lived in a windy place, and there was nothing they could do to make the wind stop. All they could do was find ways to deal with it.

By the 1990s, the growing market for wind-generated energy in the western United States had convinced some developers that the near-constant breeze blowing in southern Klickitat County was more than merely a source of irritation—it was potentially an asset of great commercial value. Developers increasingly began contemplating the possibility of capitalizing on the windiness of the lower Columbia Gorge region. From a developer's perspective, Klickitat County and the surrounding area had more to offer than just strong wind resources. Because the Bonneville Power Administration had constructed multiple hydroelectric dams along the Columbia River decades earlier, there was an unusually plentiful amount of high-voltage transmission capacity available in the area. This ready transmission infrastructure had the potential to connect new wind farms into California's lucrative electricity markets, where wind-generated power was beginning to command significant price premiums. The Columbia Gorge region's hydroelectric systems could also serve as a good complement to wind farms, allowing grid operators to better manage wind energy's intermittency challenges by adjusting nearby hydropower production. The rare combination of these transmission-related advantages and above-average wind resources made southern Klickitat County a particularly attractive potential wind farm site.

A developer named Dana Peck was among the first individuals to appreciate the wind energy potential in Klickitat County. He moved to the county in the 1990s to attempt to develop a wind farm in the county's Columbia Hills area for the company Kenetech. Peck was a pioneer of sorts for wind energy in Klickitat County—an outsider who moved into to the region and quickly recognized that wind energy seemed a natural fit for this struggling rural area.

When Peck arrived in Klickitat County, the local economy was stagnant and its fiscal challenges were growing. Farming and cattle ranching had historically been the county's predominant industries. County residents that had long relied primarily on these industries for income were finding it more and more difficult to make ends meet. Peck viewed wind energy as one of the most promising ways to revitalize the county's economy and improve its shrinking municipal budgets.

Despite the diligent efforts of Dana Peck and others, Kenetech's wind farm plans never took off the ground. Poor timing was unquestionably part of Kenetech's problem: the market demand for wind-generated energy was still too tepid to make a new wind farm project commercially viable in the region. Congress also had not yet enacted generous federal tax credit programs to promote renewables, and most of the other government programs that have spurred wind energy's growth in the past decade had not yet been put into place. The project's economics were simply not good enough to justify pursuing wind farm development at that time.

[margin note: project was unsuccessful]

[margin note: economic hurdles]

However, Dana Peck believed that market conditions and federal policies were not the only reasons that wind farms had not yet sprouted in Klickitat County. While working for Kenetech, Peck had experienced first hand the frustrations of trying to obtain approvals for wind energy development through the county's outmoded and burdensome system of conditional use permit review.[3] He felt that this arduous approval process, and the strict environmental review requirements mandated under the Washington State Environmental Policy Act (SEPA), involved so much expense and risk that few wind farm developers would ever want to pursue projects in the county.

[margin note: permit process]

Rather than leave Klickitat County in light of these problems, Dana Peck chose to stay and lead an effort to remove some of the obstacles that had prevented his company's wind farm project from taking shape. Peck worked his way into local government administration, eventually accepting a position as Klickitat County's Director of Economic Development. He then rolled up his sleeves and went to work, spearheading revisions to the county's ordinances so that the county would be better positioned to attract wind energy development dollars when market conditions improved.

A welcome mat for wind farms: the energy overlay zone

Dana Peck's years of experience as a wind energy developer with Kenetech gave him a good sense of what it would take to woo wind energy developers into Klickitat County. His policy strategy was simple: streamline the county's wind farm permitting process, eliminating as many unnecessary regulatory risks as possible. Peck was not advocating a laissez-faire, anything-goes approach to wind farm approvals. However, he did want to introduce more developer-friendly policies in the county—policies that would lay down a figurative welcome mat for responsible wind energy development.

[margin note: minimize regulatory risks]

Peck and other Klickitat County officials ultimately crafted an innovative "energy overlay zone" (EOZ) ordinance to promote wind energy in certain county areas. Using available data on wind resources, transmission availability, and wildlife habitat areas, officials first identified locations within the county that seemed most suited for wind farms. Then, officials expended municipal funds to finance an environmental review of possible wind energy development in those areas, eventually obtaining a programmatic environmental impact statement (EIS).[4] For lands situated within the boundaries outlined on the final EOZ map, this programmatic EIS satisfied many of the SEPA requirements commonly associated with wind energy projects. The ordinance, adopted in 2004, effectively converted the wind farm permitting process in those areas from a complex conditional use proceeding into a streamlined planning director decision on an administrative permit—a far simpler and less challengeable procedure.

gov took over the site iden- tification & EIS process rather than developers doing it themselves from scratch

Klickitat County's adoption of its programmatic EIS meant that proposals for wind energy development within the county's EOZ could gain many of their required approvals in as little as 45 days—far more quickly than if a wind energy developer had to commence the permitting and SEPA review process from scratch. This creative policy move sent a powerful signal to potential developers that Klickitat County was "open for business" and eager to help facilitate wind farm development within its boundaries.

Importantly, Klickitat County's investment of time and money to formulate and adopt the EOZ also helped to nurture general local support for wind energy development. This building of grassroots support occurred while the idea of a wind farm was still entirely "in the abstract"—before the county had heard any permit requests for wind farms in particular areas. The programmatic EIS and EOZ ordinance were public manifestations that a political majority in the county favored wind energy development. They fostered a sense of community-wide commitment to wind energy that proved very beneficial later when the county began entertaining specific project proposals.

If gov does it, more local input is more public support

A seasoned developer

Not long after Klickitat County adopted its EOZ ordinance, the county attracted the attention of Gerry Monkhouse and Gary Hardke, the co-owners of Cannon Power Group (Cannon). Monkhouse and Hardke had been in the renewable energy development business for more than 20 years, ranking them among the most experienced wind energy developers in the country at the time. Although Cannon's headquarters were in San Diego, California, Monkhouse and Hardke had been actively involved in developing dozens of renewable energy projects across the globe. Having a wealth of experience, they had a knack for identifying promising development sites.

In the early 2000s, Monkhouse and Hardke began exploring the possibility of developing a large wind farm in southern Klickitat County along the northern rim of the Columbia Gorge. They specifically set their sights on a large site near the town of Goldendale, Washington. The site was situated just a few miles east of a cliff that had long been known as "Windy Point" because of its perpetually blustery conditions. Convinced that this area possessed all of the characteristics of a prime location for wind energy development, Monkhouse and Hardke started making plans to pursue a potential project there.

Well aware that delays and difficulties in obtaining permitting approvals were among the greatest risks Cannon faced in these contexts, Hardke, an experienced attorney, was intrigued when Klickitat County adopted its EOZ and programmatic EIS. These unusual programs not only promised to expedite the permitting process; they also demonstrated that the local government and community at large were serious about having a wind project in their midst. By adopting the EOZ, local officials had signaled that they understood the ramifications of wind energy development and genuinely wanted to attract developers to the county.

signaled readiness to developers

The prospect of a wind farm in Klickitat County unquestionably seemed appealing to Monkhouse and Hardke at first blush. However, they knew that they needed to investigate more deeply before making any major investments toward building out a project there. Would it really be feasible for Cannon to connect a new wind farm in the area into BPA's high-voltage transmission system and deliver the project's power to California utilities? Would state or local opposition stemming from possible impacts on wildlife, views, or cultural resources prevent development, despite the existence of the EOZ? And was the community truly prepared for all of the changes that a new wind farm would bring? Monkhouse and Hardke wanted to become comfortable with all of these issues before spending too much money in active pursuit of a project.

Cooperating with rivals to connect into the grid

A sure means of delivering electricity to end users is a necessary requirement for any wind energy project, so transmission issues loomed large as Cannon first mulled the possibilities of development in Klickitat County. There was plenty of excess capacity on BPA's nearby 500 kV transmission system to deliver additional power to California, but new lines and facilities would undoubtedly be needed to connect that system to Cannon's potential project site. Such facilities could have a hefty price tag and would require significant cooperation with the local utility district.

Over the course of several months, Cannon engaged in discussions with BPA and with the Klickitat Public Utility District (KPUD) regarding potential expansions of the county's transmission infrastructure to facilitate wind energy development. As these discussions progressed, it soon became obvious that Cannon could not foot the entire bill for these transmission expansions on its own. Accordingly, what began as a bilateral discussion between Cannon and KPUD quickly transitioned into a complex multilateral negotiation involving multiple other wind energy developers who contemplated developing their own, competing projects in the area.

This sort of bargaining dynamic—with several competing developers trying to pool resources to site and fund transmission facilities that are critical to each of their rival projects—inherently lends itself to tension. On the one hand, Cannon and the other developers in these negotiations recognized that some degree of cooperation was imperative to their success; if they couldn't reach agreement, it was quite possible that *all* of their proposed projects would fail. On the other hand, each developer at the negotiating table was also fiercely competing with everyone else to secure wind energy leases with landowners and to get community members to coalesce around its own particular project. Due to practical limitations, it was more than likely that some of these developers' proposed projects would ultimately fail.

[margin note: Coordination w/ other developers]

Comprehending the nature and implications of this delicate dynamic, Cannon tried to help this group of rival developers focus on the greater good that could be gained from coordinated effort. Rather than bristle at the notion of working with direct competitors to help fund transmission, the company and its lawyers sought to guide funding discussions toward workable solutions. For the most part, the developers in the group maintained an air of congeniality, listening to the views of all developers that were serious about participating.

Eventually, a tentative deal was reached: Cannon and a handful of other wind energy developers agreed in principle to collectively fund the $35 million that KPUD would need to build a new substation and transmission connections. In return, each developer would receive interconnection capacity at the new substation in an amount that was roughly proportional to the size of its respective financial contribution to the transmission project.

By overcoming its differences, this unlikely group of competing wind energy developers was able to create grid connection opportunities that would have otherwise been unavailable. The final arrangement was far from ideal, but it was still more than adequate from Cannon's perspective. For $16 million, the company could secure 500 MW of interconnection capacity—enough to connect a $1 billion wind farm to California electricity markets 800 miles away.

Securing lease rights

Satisfied with the quality of wind resources and transmission availability in Klickitat County's Windy Point area, Cannon next turned its attention to assessing the one other critical element for successful wind energy projects: a group of landowners who are ready and willing to lease their land. Monkhouse and Hardke sincerely believed that a wind farm would create net benefits for the area, but could they persuade local residents to share their vision? They determined that the only way to get an accurate perspective on that question was to make a series of trips from California up to Klickitat County and talk face-to-face with the farmers and ranchers who lived there.

Local government officials in Klickitat County had a good sense of the potential impacts of having commercial-scale wind farm nearby, but many of the other locals were far less familiar with these projects. Consequently, Monkhouse and Hardke spent dozens of hours chatting with local residents in homes and coffee shops in and around Goldendale trying to educate community members about wind energy. Residents raised all sorts of questions about Cannon's proposed project and how it might affect the community:

"How will the turbines affect my ranch?"
"What exactly will these windmills look like?"
"How loud will they be?"
"Will I still be able to hunt on my property after the turbines go in?"
"Why should I believe that Cannon can pull this off, after so many other developers in the past have failed here?"

One by one, group after group, Hardke and Monkhouse candidly responded to these numerous inquiries. Rather than giving sugar-coated answers, they endeavored to be as honest and straightforward as possible. Having developed wind energy projects many times before, Cannon knew that the worst thing that could happen would be for residents to be unpleasantly surprised when new roads, transmission lines, and turbines started popping up around them.

At the same time, Cannon tried to emphasize to landowners the potential benefits of entering into a wind lease. The company noted that the new fences, roads, cattle guards, and wells installed in connection with its wind farm could actually enhance property values and aid farmers and ranchers. Some ranchers have even discovered that their cattle like to lounge in the shade of turbine towers when the hot sun beats down on long summer days. And then there were the financial benefits of signing a wind lease: what rational landowner wouldn't want to lock in 20 years or more of significant additional income in exchange for some very minimal disruptions to existing farm or ranch operations?

The conversion of Ruth Davenport

Although Cannon's proselytizing resonated with several locals in the Windy Point area, not everyone was easily convinced. The Davenport family was one influential and tight-knit group that was initially skeptical about Cannon's plans. The family's land holdings made up a large portion of Cannon's desired project area. They had been ranching on the land for five generations and were well known to nearly everyone in the Goldendale area.

Typical of even large-scale ranchers in the county, the Davenports were of relatively modest means. The matriarch of the family was Ruth Davenport, a great-grandmother who was well into her 70s in age. Ruth had seen first hand the gradual decline of Klickitat County's local economy over her four decades of living there. The local aluminum plant had shut down a few years earlier. Changes in the cattle industry were placing ever more financial strain on ranches in the region. Economic opportunities were very limited, and local jobs were difficult to find. During her time as president of the local school board, Ruth had watched with sadness year after year as many of the community's brightest students were forced to leave the county to find employment. She initially doubted that wind energy could truly help address these chronic economic woes—conditions that even affected her own day-to-day life. Despite her age, Ruth worked at the local general store to help cover family expenses.

One morning, early on in Cannon's exploration of a possible wind farm project in Klickitat County, Monkhouse and Hardke met Ruth Davenport and others for breakfast. In the midst of this conversation, Ruth began delineating her community's struggles. She seemed to hold little hope for improvement, mentioning the closed aluminum plant, the floundering ranches, and the dearth of jobs. Her expressions of frustration culminated as she described how her granddaughter, Brandy, had to accept a job more than 100 miles away so that she could make use of her accounting degree. Tears welled in Ruth's eyes. Monkhouse and Hardke sat silent and still, not sure what to say. Then, Monkhouse broke the silence.

"Ruth," he said, "I'll make you a promise … if you work with us, we're going to make this wind project happen. And when we do, we'll bring jobs so that local kids can stay in the county."

Ruth believed Monkhouse and eventually signed a wind energy lease with Cannon. Other landowners in her family did the same.

Monkhouse wisely kept his promise to Ruth. One of the first individuals that Cannon hired in connection with its Windy Point/Windy Flats project was Ruth's granddaughter, Brandy. Brandy's new position with Cannon allowed her to move back to Goldendale, where she began working in a small office space that Cannon had leased on the edge of town. Her new job meant that she could visit Ruth regularly and once again spend quality time with her grandmother and other family members. Ruth could not have been happier. The income Ruth received under her wind energy lease with Cannon greatly improved her family's finances. Even better, there were new signs of optimism in Goldendale—the remote community that had long been her home.

Kurt Humphrey and the courting of Eleanor Dooley

Over the course of several months, Hardke and Monkhouse grew increasingly satisfied that they could successfully develop a wind farm project in Klickitat County. However, because they were "executive types" with offices in California, they knew that they were relatively out of touch with the day-to-day life of the county's rural farmers and ranchers. To succeed, they would need somebody on the ground in the county every day working as an advocate and liaison for Cannon and its proposed project. This liaison would obviously free up time for Hardke and Monkhouse to work on other aspects of the project back in San Diego. Hopefully, this person would also be better able to relate to the local residents and would not be perceived as a stuffy outsider. Cannon began searching for a candidate with enough know-how and local charm to build goodwill and secure the leases, easements, and other interests the company would need.

Cannon found its man in Kurt Humphrey, a down-to-earth individual with a slight western drawl and a folksy demeanor. Humphrey lived almost two hours away on the outskirts of Portland, Oregon, but had spent much of his career negotiating oil and gas leases in rural areas. An avid outdoorsman, Humphrey's typical attire consisted of jeans, a flannel buttoned shirt, leather boots, and a cowboy hat. His unassuming appearance and rural background enabled him to fit right in with the people of Goldendale and the surrounding area.

In the ensuing years, Humphrey spent so much time in and around Goldendale that he came to regard it as a second home. As the project progressed, Humphrey grew more and more sympathetic of locals' interests and concerns, so much so that at some points Hardke and Monkhouse jokingly accused him of "going local" at Cannon's expense. In reality, Humphrey's visible concern for the people of Klickitat County was among the greatest assets he brought to Cannon. His outspoken commitment to "doing right" by landowners allowed him to get crucial wind leases signed when no one else could.

Humphrey's signature accomplishment at Windy Point/Windy Flats was his winning over of Eleanor Dooley, an elderly widow who owned vast acreages of land in southern Klickitat County. When it came to wind farm development, Dooley was a notoriously tough nut to crack. She had rejected countless offers from wind energy developers to lease her land over the years, declaring on multiple occasions that she did not want any part of "the windmills." Community members had warned Cannon about Dooley's staunch opposition to wind energy, suggesting that attempts to lease her land were hopeless and weren't worth the effort. Dooley was nearly 90 years old and was deeply set in her ways. She had no compelling financial needs that would motivate her to allow development on her land.

Unfortunately for Cannon, Dooley held title to large tracts of land situated squarely in the middle of the company's planned wind farm. The commercial viability of its project was far less certain without lease rights covering Dooley's several parcels. Somehow, Cannon had to get Dooley on their side—a seemingly impossible task. The fate of Cannon's entire project could depend on it. How could Cannon persuade this shrewd, petulant frontier woman to sign at the dotted line? This formidable challenge fell to

Humphrey, who was instructed to give top priority to getting Dooley on board.

At first, Humphrey's efforts to befriend Dooley got him nowhere. She refused to even give him the time of day. How could Humphrey win Dooley over if she wouldn't even chat with him? He knew that he had to be more creative if he were to have any chance of succeeding. Humphrey started dropping by Dooley's home with small, insignificant gifts and trying to get just a few words in at the doorstep. Some attempts were more successful than others, but Humphrey persevered.

Then, somewhat to Humphrey's surprise, Dooley started to soften. When Humphrey stopped by on wintry days, she began inviting him to come inside her humble ranch home to get out of the wind and cold. There, Humphrey would sit and listen to Dooley tell stories of "the way it was" in Goldendale decades ago and how the community had changed. Humphrey complimented her on her garden. He asked her about her family. He spent quality time with her and gradually became her friend.

Somewhere along the line, Dooley became convinced that Humphrey was not visiting her purely so that he could induce her to execute a wind energy lease with Cannon. It was at this point that she expressed a willingness to sign.

Dooley ultimately became the single biggest landowner involved in Cannon's wind farm project. Humphrey continued to visit her often, even long after wind turbines had sprouted up on her land. Dooley passed away in 2013 at the age of 97, but her legacy and name live on in Goldendale. As an expression of gratitude, Cannon named one of the project's transmission substations in her honor.

Building community support

As difficult as it was to get some landowners to execute wind energy leases, Cannon knew the greatest opposition it would face in trying to develop a wind farm in Klickitat County was not likely to come from its own lessors. Once these landowners signed wind energy leases with Cannon, they generally became ardent supporters of the project and pushed for its speedy completion so that they could begin collecting lease income. Instead, the most difficult opposition Cannon was likely to face was destined to come from neighbors and other outsiders—individuals and groups who feared that Cannon's project would not create any net benefits for them. These outsiders had no prospect of receiving monthly rent checks under wind energy leases and thus had far fewer reasons to support Cannon in its goals.

In Klickitat County, neighbors and other citizens vastly outnumbered the small group of landowners who would be signing leases with Cannon. Without adequate support from these outsiders, Cannon's project stood little chance of getting off of the drawing board. What could Cannon do to persuade these citizens residing *outside* the bounds of its development area that its proposed project was in their interest as well?

Cannon had learned from prior experience not to underestimate the importance of neighbors and outside groups when developing a wind farm. The company's strategy for gaining and keeping this contingent's favor was threefold: (i) be as transparent and open as possible, (ii) hire locally whenever feasible, and (iii) give generously back to the community through charitable and other donations.

Communication and transparency

Transparency and proactive communication with locals had spared Cannon from public relations challenges on multiple occasions in connection with other renewable energy developments, so Cannon was committed to these principles in connection with its Klickitat County project. Throughout the project's development Cannon regularly released photographs and short articles describing its progress. Over time, local residents increasingly followed these updates to stay informed. Although Cannon had to spend precious time and money on this sort of outreach, its investments paid valuable dividends at certain points in the process.

One incident that highlighted the value of Cannon's public relations efforts in Klickitat County involved the global financial crisis that erupted in late 2008. Cannon was actively preparing for a large financing closing on a major project phase when the crisis broke. Within days, a handful of news outlets began reporting that Cannon's lead lender, HSH Nordbank AG, could be on the brink of failure. This unwelcome news caused rumors to spread like wildfire through the tiny town of Goldendale. Was Cannon's chief lender really about to fail? And if so, would the bank's failure doom further development at Windy Point?

Cannon was understandably afraid that this gossip might deter some landowners from cooperating or cause them to commence discussions with competing developers. Fortunately, its communications plan meant that the company had a system in place to swiftly dispel the rumors that were swirling around its project. Cannon posted statements allaying local fears about the fate of its financing and was thereby able to weather what had rapidly erupted into a dangerous public relations storm.

Hiring locals

Cannon also sought to build favor within Klickitat County by injecting money into its local economy through the hiring of local contractors and personnel whenever possible. Obviously, the project's turbines could not be manufactured in the county. However, much of the work associated with building and grading roads, surveying the land, pouring concrete, digging wells, laying and burying low-voltage transmission wires, and other aspects of the project was capable of being done by local companies. Rather than bring in California companies to handle the wind farm's construction, Cannon deliberately sought out local contractors and workers for these tasks. Cannon even hired locals for its key administrative positions. Danielle and Keelie Olsen—two sisters who were longtime friends of Brandy Davenport—began working for Cannon relatively early in the wind farm's development and have continued handling administrative duties for Cannon even long after completion of the project.

Cannon's commitment to hiring locally led to the creation of more than 150 local jobs and the infusion of tens of millions of dollars into Klickitat County's economy, all of which helped to buttress community support for Cannon's wind farm. Even though most Goldendale residents did not own land within the project area and thus received no lease payments, most of these residents did see positive economic impacts from the project in their own lives. A letter from Goldendale's mayor, Arletta Parton, to Cannon during the development process highlighted some of these indirect benefits:

> I look at the motels being filled, restaurants being busy, the RV park being full … thanks to all of you for the overall benefits you have brought to my community.[5]

Although Cannon's practice of hiring locally when possible sometimes involved a bit of extra up-front cost because of additional training time, the increased community support gained from this approach more than justified these expenses in the end.

Spreading donations around

The other primary strategy that helped Cannon gain and maintain local support for its wind farm project was to make modest donations to a long list of community groups. One of Brandy Davenport's duties as an employee of Cannon was to identify local organizations that she felt were deserving of charitable assistance. Rather than making a small handful of larger donations, Cannon tried to spread its donations as much as possible so that they would directly benefit a larger proportion of the community's residents. The company donated to between 10 and 20 different organizations per year during project construction.

Some of Cannon's donations went to predictable recipients—organizations such as the local food bank or the school district. However, the company also made donations to dozens of groups that rarely receive funds from developers. The company donated to the Klickitat County Livestock Growers, the Goldendale High School Rodeo Team, the Goldendale Police Department, Ducks Unlimited, and the Centerville Church, to name a few.

This strategy of distributing donations among a numerous and diverse spectrum of interests helped Cannon to forge goodwill among a wider range of the local populace.

Partnering with Maryhill Museum of Art

The most notable set of donations that Cannon made while developing its wind farm in Klickitat County went to the Maryhill Museum of Art (Maryhill)—an eccentric art museum nestled on a hill just above the Columbia River Gorge. Maryhill's 5,300 acres of land were situated immediately adjacent to Eleanor Dooley's property. Much of this acreage was rich with prime wind energy resources.

Initially, there were reasons to fear that Maryhill could be among one of the greatest opponents to Cannon's plans for a wind farm in the area. Because of the museum's location, some of the turbines Cannon planned to install would almost certainly be visible from the museum grounds. Maryhill's pristine rural setting was part of what made it unique. Few museums in its position would want such a backdrop to be cluttered with towering wind turbines. And yet, Cannon knew that Maryhill would need to be in its camp if its wind farm project was going to proceed.

Fortunately for Cannon, Maryhill was not a conventional art museum. Its eclectic collection comprised of classical European paintings alongside chess sets, and religious and native art was itself an anomaly in rural Washington. Maryhill's expansive grounds also featured a replica of the Stonehenge monument and a peculiar stretch of steep, winding roads that annually served as the venue for an international skateboarding competition. In part because of its unusual location and characteristics, Maryhill also needed cash. Due to rising gasoline prices and other factors, the number of paying visitors to the museum had been declining for nearly a decade.

Maryhill's avant-garde identity and need for additional revenue made it a natural partner for Cannon, who vigorously worked to establish a positive rapport with museum officials. These efforts paid off: Maryhill ultimately concluded that wind energy development was fully consistent with its mission and vision for the future. The museum became another one of the major landowners involved in Windy Point/Windy Flats, allowing Cannon to erect more than a dozen conspicuous wind turbines on the museum's land. Today, Maryhill receives roughly $200,000 per year in lease payments under its wind energy lease with Cannon. This source of steady revenue and more than $500,000 in charitable contributions from Cannon gave Maryhill the confidence to undertake a $10 million addition to its facilities—a dramatic expansion made possible only because the museum now supplements its income by harvesting its wind.

Resolving neighbor concerns

Cannon's public relations efforts, local hiring practices, and charitable donations to local groups certainly helped to foster positive feelings about the company and its project in Klickitat County. Nonetheless, neighbors and local advocacy organizations still raised multiple concerns about the project during the permitting and development process.

Even though Cannon's project site was within Klickitat County's EOZ area, Cannon still had to obtain at least 20 different permits in connection with its project. Merely satisfying the lengthy list of conditions associated with these permits enabled Cannon to prevent the vast majority of potential neighbor conflicts from its wind farm, including such issues as dust, noise, erosion, stormwater drainage problems, fire hazards, and electromagnetic interference. However, even with all of these onerous conditions in place, some neighbors and other stakeholders were still worried about how Cannon's project might adversely affect them.

Ice throws

One concern that emerged late in the course of permitting for the project was the possibility that turbines might throw chunks of ice onto U.S. Highway 97, a public road that ran north–south through the center of the project. As described in Chapter 2 of this book, in cold, wet conditions it is possible for turbines to become partially covered with snow or ice. As a turbine's blades rotate, that motion can fling ice fragments through the air, sometimes throwing them hundreds of meters away from turbine towers. Given Klickitat County's frigid winter temperatures and the proximity of some of Cannon's proposed turbine sites to the highway, ice throwing posed a genuine risk in connection with the project.

Klickitat County officials somehow overlooked this ice throwing risk when they drafted its EOZ ordinance, so when Cannon proposed turbine sites near U.S. Highway 97 the ordinance was silent on this issue. When the potential for ice throwing drew attention during the permitting of those sites, Cannon agreed to adjust some turbine locations to help minimize ice throwing risks and also proposed an "ice throw protocol" that the County ultimately incorporated as a new permit condition. Under the protocol, Cannon agreed to inspect and potentially shut down turbines whenever low temperatures and precipitation or high humidity created conditions capable of generating ice throw risks. This combination of Cannon's protocol and careful siting has proven to be quite effective: as of late 2013, there had been no reported damage from ice throws at the project.

View impacts

The other main worry that neighbors raised in connection with Cannon's project at Windy Point/Windy Flats was that the project's towering turbines might tarnish territorial views in and around the Columbia Gorge. Maryhill initially expressed concern about impacts on views from its property, and other landowners in the area had similar reservations.

Cannon's primary strategy for allaying these fears was to prepare highly accurate view simulation photos and share them with landowners in the area. These computer-generated images, which showed precisely how the turbines would appear once installed, did manage to put many landowners at ease. In the few situations where discontent persisted even after showing simulation photos, Cannon rearranged turbine sites to ameliorate the problem.

Unfortunately for Cannon, even these simulations and minor adjustments to turbine sites were not enough to satisfy one advocacy group residing in the area. The greatest opposition that Cannon faced with respect to the viewshed impacts of its project came from Friends of the Columbia Gorge (FCG). FCG is a nonprofit organization that was founded in 1980 and is based in nearby Portland, Oregon. The group's stated mission is to "vigorously protect the scenic, natural, cultural, and recreational resources of the Columbia River Gorge."[6] The Gorge serves as a popular recreation area for residents of the region, providing unique opportunities for windsurfing, kiteboarding, fishing, river sailing, and other activities.

When Cannon received a permit for one of the latter phases of its wind farm project that included several turbine sites near the Gorge's north rim, FCG swiftly filed an administrative appeal. The group's primary argument was that some of the turbines authorized under the permit would degrade and tarnish views from within the Gorge. Hoping to prevent this appeal from further delaying project construction, Cannon prepared a very large number of additional simulation images showing how the turbines would appear from various Gorge locations. These simulations did not placate FCG, and the group continued to argue against some of Cannon's proposed turbine locations.

Cannon ultimately reached a settlement with FCG that required Cannon to eliminate of some of the phase's more visible turbine sites and to move other sites further away from the rim's edge. Because of their elevation and proximity to the Gorge, some of Cannon's forfeited sites had very favorable wind resources. Although the terms of the settlement were costly for Cannon, the agreement enabled development to move forward without further delays—a critical benefit at that stage of the project.

Competing with other developers

Not surprisingly, neighbors and special interest groups were not the only potential opponents of the company's proposed wind farm. Several competing developers were actively seeking to build their own wind energy projects in Klickitat County and viewed Cannon as a direct threat to their plans. Assembling a critical mass of leased parcels is vital to the success of any wind farm, and there was not enough developable land in the county to satisfy the appetites of all of the companies that were pursuing projects there. Consequently, Cannon found itself embroiled in a merciless battle against multiple other developers to secure wind energy leases.

Competing for leases

Cannon was well aware that most of the landowners it spoke with about the possibility of entering into wind energy leases were also hearing sales pitches from competing developers. Hoping to distinguish itself from the crowd, Cannon thus seized on every opportunity to highlight what it perceived to be its unique strengths. Cannon was much smaller than several of the other companies pursuing wind farms in the county. Many of Cannon's competing developers were subsidiaries of large, publicly traded companies with recognizable names. Aware of this difference, Cannon portrayed its diminutive size as an advantage, touting that landowners who signed leases with the company would be able to work directly with the "owners" of the company. Cannon argued that this was far more desirable than working with a junior development agent of a massive, impersonal corporation who was less likely to care about landowner needs once a lease is signed.

Cannon also sought to depict its style of wind energy development as one that would *benefit* rather than *hinder* landowners' existing farming and ranching operations. For instance, Cannon paved dozens of miles of previously unimproved dirt roads within the project area and created a network of new roads, greatly expanding landowners' vehicular access to their land. The company also dug multiple new water wells to assist with dust suppression during construction, and these wells became valuable assets to landowners after construction was done. Cannon even installed numerous new cattle guards and lockable gates on roadways to help prevent livestock from escaping.

Although these strategies surely helped Cannon in its battle for signed wind energy leases in Klickitat County, the company gained its greatest comparative advantage by offering very favorable, "above market" lease terms to landowners. Hardke and Monkhouse had come to espouse this strategy of offering generous lease terms based on their experiences in developing other wind energy projects over the years. To be sure, the relatively high up-front lease payments and percentage rents in Cannon's wind energy leases cost the company millions of dollars and contrasted sharply with the growing trend of nickel-and-diming in the industry. Nonetheless, Hardke, who had negotiated well over 100 wind energy leases at that point in his career, was a firm believer in this costly approach.

In the minds of Hardke and Monkhouse, the benefits of offering above-market lease terms were at least twofold. First, this approach was undeniably helpful at the lease acquisition stage, persuading some landowners to sign wind energy leases with Cannon rather than with competitors. However, this strategy also paid some less-obvious dividends as project development progressed. Hardke and Monkhouse knew that the 20-year relationship forged between Cannon and its project landowners under these leases would inevitably encounter some rocky moments: construction delays, last-second document signings for financing closings, or other occasional challenges were bound to arise. Generously compensating project landowners made these lease relationships feel more like partnerships, promoting greater landowner cooperation and patience when Cannon needed it most.

One other obstacle that Cannon faced in its leasing efforts was the potential challenge of negotiating separate wind energy lease agreements with more than 20 different landowners. The sheer number of lease agreements was itself daunting: if variations were allowed in each one, how would Cannon or its lawyers keep track of all of them? And how could Cannon provide adequate assurance to each landowner that the specific terms in its own lease were just as favorable as those in the several other leases involved in the project?

The complexity of Cannon's wind energy lease agreement itself threatened to make the leasing process difficult. To pass muster with Cannon's large institutional lenders, the form agreement was quite lengthy and contained several pages of "legalese" that was barely decipherable for many landowners. Because most of the small-town lawyers who were likely to represent landowners in these lease negotiations had minimal experience in wind energy leasing, even they were likely to be intimidated or confused by some provisions in the document. In the face of these difficulties, how could Cannon persuade each of its several project landowners and their attorneys that its lease agreements were reasonable and fair?

Cannon stumbled upon an ingenious solution to this predicament while negotiating its lease with Goldendale Aluminum—another owner of some of the windiest developable land in Klickitat County. Goldendale Aluminum had operated an aluminum smelting plant on land near the Gorge for decades. The plant had fallen onto hard times and shut down a few years earlier. When Cannon approached Goldendale Aluminum about the possibility of leasing its land for a wind farm, the company asked Brett Wilcox to negotiate a wind energy lease on its behalf.

Wilcox was a sophisticated, experienced attorney and businessman who had studied law at Stanford and had a strong reputation in Klickitat County. Aware that county residents greatly respected the judgment of Wilcox and Goldendale Aluminum, Cannon devised a plan to help ease its leasing process for landowners. First, Cannon negotiated long and hard with Wilcox on Goldendale Aluminum's wind energy lease, ultimately agreeing to a reasonable lease document that was satisfactory to both parties. Cannon then approached its other potential landowners and offered the Goldendale Aluminum lease as a template for their leases. "If this agreement is good enough for Goldendale Aluminum and Wilcox," Cannon argued, "you can rest assured that it is good enough for you, too."

Cannon's strategy worked brilliantly. The majority of the project's other landowners and their attorneys challenged relatively few lease provisions. Their relative deference to Cannon's template agreement was partly attributable to their knowledge that a highly-skilled lawyer had already done significant lessor-side negotiating and that those efforts were reflected in the template. Because of this, Cannon was able to spend far less time and lawyer fees than it might have anticipated to secure most of its project leases in relatively short order. The provisions contained in Cannon's final collection of signed leases were also more uniform than they would have otherwise been.

Competing for wind

The other risk that competing developers posed for Cannon was the possibility that wind turbine wake interference conflicts like those described in Chapter 3 could arise. Like many jurisdictions, Klickitat County had ordinances that allowed developers to site turbines relatively close to parcel boundary lines. The county's safety-based setbacks did not take turbine wake interference into account and thus created a potential for wake conflicts with neighboring developers.

Cannon encountered the problem of wind turbine wake interference on at least one occasion in connection with the Windy Point/Windy Flats project. The company had leased a tract of land just east of Highway 97, near the center of its project, on which it intended to site multiple turbines. Cannon then learned that a competitor operating under the name of Northwest Wind Partners (NWP)—a subsidiary of what is now EDF Renewable Energy—had leased an adjacent tract where it planned to site several turbines of its own. If Cannon and NWP each installed turbines in desired locations on their respective parcels, a wake created by one of Cannon's turbines would reduce the wind energy productivity of an NWP turbine, costing the NWP and its lessor tens of thousands of dollars.

When NWP learned about Cannon's intention to install a turbine immediately upwind of its proposed turbine site, the company contacted Cannon and threatened to sue the company for nuisance and under other theories. NWP demanded that Cannon relocate its turbine site far enough away from the property boundary line to avoid wake interference problems.

Unsure how to respond to NWP's demands, Cannon contacted its legal outside counsel and asked for advice. What was the likelihood that NWP would prevail in such a dispute? Was a court likely to hold that Cannon's operation of an ordinary wind turbine in compliance with applicable setbacks constituted an actionable nuisance under relevant law? Or was

Cannon entitled to capture wind currents above its leased parcel, regardless of its impacts on NWP's wind farm operations downwind?

Cannon's attorneys searched and searched under applicable law and found no definitive answer to their client's questions. The law regarding turbine wake interference was totally unclear. Rather than take its chances in court, Cannon elected to approach NWP and try to negotiate some sort of mutually beneficial settlement with the company.

Ultimately, to the benefit of both parties, cool heads prevailed. The parties managed to finalize and execute a settlement agreement resolving the wake interference issue. Their agreement described, in metes and bounds, an "Upwind Prevailing Wind Direction Exclusion Zone" within which each party agreed not to install any turbines. The agreement also involved the exchange of a transmission line easement and some other land rights. After executing the agreement, the parties recorded a memorandum of the agreement in the county real estate records. The agreement was a surprisingly good outcome for Cannon and for NWP—it provided certainty to both parties on an issue for which there remained plenty of uncertainty under state law.

Addressing wildlife conflicts

From the very beginning, one favorable aspect of south-central Klickitat County from Cannon's perspective was that it contained relatively few sensitive habitat areas that might complicate efforts to build a wind farm there. Private landowners had been using most of the area for farming and ranching activities for several decades, so additional development on it was unlikely to trigger major new conflicts with wildlife. Nonetheless, Cannon knew that some limited wildlife-related issues could arise. The project site did include a few pockets of potentially sensitive habitats that were likely to require special attention. Some aspects of commercial wind farms—including their very tall towers and large moving parts—are also inherently more likely to disrupt certain wildlife than ranching and farming.

Habitat mitigation leases

Many of Cannon's steps to mitigate the wildlife impacts of its project resulted from permit conditions negotiated during the state and local approval process. As is commonly the case, these conditions required Cannon to do more than merely conduct environmental studies and seek to site turbines, roads, and other improvements away from sensitive areas. Most notably, Cannon entered into two long-term habitat conservation leases as a form of partial off-site mitigation for its project's anticipated impacts on local animal species.

Cannon's first habitat conservation lease set aside 24 acres of land as a wildlife preservation area, precluding any development within the leased premises. The Washington Department of Fish and Wildlife selected a site owned by a local rancher as the preserve site for this lease because of its high-quality Oregon white oak and other habitat values. The lease obligates Cannon to pay $10,000 per year to the site landowner as rent—a total of $240,000 over the lease's 20-year term.

In connection with permitting for a later project phase, Cannon executed a second habitat conservation lease that preserves an additional 30 acres of land selected by Klickitat County. This second lease, which covers certain high-quality, spring-fed riparian habitat, also obligates Cannon to pay upwards of $200,000 in total rent over 20 years.

In addition to these commitments, Cannon's project permits require the company to engage in specific ongoing efforts to monitor and help prevent harms to wildlife throughout the life of its project. For instance, Cannon was obligated to conduct "post-construction fatality monitoring"—the counting of bird kills within the project area—for one to two years after the project was up and running to help ensure that its anticipated impacts on birds were consistent with actual impacts. The company also must implement a Wildlife Incident Reporting and Handling System that requires it to report bird and bat fatality numbers annually to government agencies. Cannon must notify those agencies within 24 hours of discovering the fatality of any threatened or endangered species on the project site. Fortunately, as of late 2013, there had been no known fatalities to such species at the wind farm.

A legal challenge over avian impacts

Even Cannon's long list of actions to protect wildlife were not enough to convince some conservation groups that wildlife preservation concerns were adequately addressed in connection with the company's project. In particular, a collection of local bird conservationists did not believe that county and state permitting requirements would adequately protect raptors and other species in and around the project site. This group had actively opposed the Klickitat County's adoption of its EOZ ordinance years earlier. Despite reports by leading avian experts predicting that Cannon's wind farm would have minimal impacts on the region's bird populations, these activists were bent on opposing the project. When the county issued Cannon its first EOZ permit in 2006, the group quickly filed an administrative appeal.

A hearing officer who heard the group's appeal ruled in Cannon's favor, but Cannon could tell that this group of opponents was not going to go away lightly. Cannon was legitimately concerned that the group's further appeals could impede progress on the wind farm. Recognizing that lengthy delays resulting from such appeals could prove extremely costly, Cannon decided to focus on trying to swiftly reach a compromise with the group so that development could proceed. Ultimately, Cannon negotiated a settlement with the group under which Cannon formed a habitat conservation fund known as the Columbia Hills Conservation Fund. As of late 2013, Cannon had contributed roughly $1.1 million to this fund. The Columbia Land Trust serves as an institutional trustee for these monies, leveraging them to acquire conservation easements in and near the Columbia Gorge that are strategically aimed at protecting bird habitats in the region.

Protecting cultural resources

Like much of the western United States, the lands involved in Cannon's wind farm at Windy Point were once occupied by a Native American tribe. The primary tribe in Cannon's project area was the Yakama Nation. Klickitat County itself is named after the "Klikitat" people, a federally recognized tribe that counts itself as part of the Yakama Nation. Although the Yakama possesses reservation lands in northern Klickitat County, the Yakamas' ancestral lands encompass the entire county, including Cannon's entire project site.[7] It appears likely that one of the Nation's former seasonal migration routes crossed through portions of the project's lands. Arrowheads and other small artifacts likely resided along these routes.

Early in the development process, Cannon wisely reached out to the Yakama Nation, seeking their input in connection with its wind farm project. However, the Nation initially showed little interest in engaging with the topic. During the construction process, the Davenport family—one of the project's major landowners—made arrangements to allow some tribal members onto its property to gather roots and other natural plant items. However, Cannon had minimal involvement in those discussions.

Although the Yakama Nation initially seemed almost indifferent to the idea of a wind farm in southern Klickitat County, the Nation gradually grew more and more outspoken and involved as construction progressed. In particular, the Nation put growing pressure on the Washington State Department of Archeological and Historic Preservation (DAHP) to monitor Cannon's cultural resource protection efforts. Over the years, the DAHP itself had grown more and more willing to assert its authority to compel projects to avoid construction in areas it deemed to have cultural "significance." In one instance at a different wind farm project in the county, the DAHP even tried to claim that an old tin can was a potentially significant cultural resource and sought to require a developer to obtain a DAHP permit to disturb it.

Fearful that unpredictable DAHP challenges could postpone the construction of its wind farm, Cannon implemented an aggressive plan to counter this risk. In particular, Cannon conducted cultural resource "site clearing" studies as early as possible on as much of its project area as it could. This strategy allowed Cannon to quickly modify construction plans to utilize less-sensitive areas when unexpected DAHP challenges arose. Thanks to this approach, Cannon was able to navigate through cultural resource concerns throughout the development of its wind farm at Windy Point with relatively few hang-ups.

The power of perseverance

Without question, there were points along the path to Cannon's completion of the Windy Point/Windy Flats project when the viability of the project was in question. Multiple other developers had previously tried and failed to build wind farms in the area. Court challenges, transmission issues, pressure from competing developers, a global economic crisis, and several other obstacles stood in the way of the project's success at various times in its planning and development stages. It would have surely been easier for Hardke and Monkhouse to pursue an ordinary residential or retail development project than to face the litany of struggles inherent in developing a wind farm. However, rather than fold in the midst of these challenges, Cannon and countless others connected to its project pressed forward and worked tirelessly to overcome them.

Today, the fruits of these labors are abundantly evident in Klickitat County. Cannon's 26-mile-long Windy Point/Windy Flats project is now in full operation. Its 500 MW of generating capacity ranks it among the largest wind energy projects in the United States. The wind farm produces enough power on average to meet the electricity needs of approximately 250,000 homes.[8] It also generates millions of dollars in lease payments annually for project landowners and millions more dollars in annual tax revenues for the county. By nearly every measure, the project has created a net benefit for the community and state where it resides.

Successful projects like Cannon's wind farm at Windy Point are a useful reminder that renewable energy development often *is* possible and worth pursuing despite the inevitable difficulty that it might entail. As government subsidies and other policies driving the demand for renewable energy gradually dry up in the coming years, these projects will face entirely new types of challenges. However, they will continue to be within the reach of determined developers in places where there is strong community support. Like many industries, renewable energy development is one that handsomely rewards persistence and hard work. And, in the case of a large wind farm, the rewards can be truly amazing.

Notes

1 The Chinook winds occur naturally throughout the Pacific Northwest. To read more about them, *see generally* Ward Cameron, *Learn About the Famous Chinook Winds* (2005), available at www.mountainnature.com/climate/Chinook. htm (last visited Nov. 19, 2013).

2 The author selected the Windy Point/Windy Flats project as the focus of the case study in this chapter because he was one of many attorneys who served as outside legal counsel to Cannon Power Corporation, the developer of the

Windy Point/Windy Flats wind farm, on matters relating to th
practicing with the law firm of K&L Gates LLP. Most of the m
in this chapter was gathered through e-mail correspondence witl
president Gary Hardke and K&L Gates attorney Brian Knox. Sc
originated from informal interviews with other parties involvec
from the author's own experiences while working on the projec.

3 For a general description of Klickitat County's transition from conditionaɪ ᵤ
   permit review for wind farms to its present Energy Overlay Zone scheme, *see
   generally* Keith Hirokawa, *Local Planning for Wind Power: Using Programmatic
   Environmental Impact Review to Facilitate Development*, 33 ZONING AND
   PLANNING LAW REPORT 1, 4–5 (2010).

4 *See* id.

5 *See* Troy Rule, *Federal Stimulus Success at Windy Point*, SEATTLE TIMES
   (May 10, 2010), available at www.cannonpowergroup.com/media/news-articles/
   federal-stimulus-success-at-windy-point/ (last visited Dec. 6, 2013).

6 Friends of the Gorge website, available at www.gorgefriends.org/section.
   php?id=9 (last visited Dec. 9, 2013).

7 For a map showing the Yakama Nation's reservation lands and ancestral lands,
   visit the tribe's website at www.ynwildlife.org/tribalhunting.php (last visited Dec.
   10, 2013).

8 *See* Cannon Power Group website, www.cannonpowergroup.com/wind/projects/
   wp-wf/ (last visited Dec. 10, 2013).

9 The role of innovation

In spite of the long list of obstacles to renewable energy development described in this book, the industry's future is exceptionally promising. One reason to be optimistic about the long-term trajectory of the renewable energy sector is the likelihood that innovation will ultimately mitigate many of its present challenges. Technological innovations are making an ever larger proportion of the planet's wind and solar energy resources available for development and are enabling more efficient storage of the energy in those resources over time. Policy innovation is likewise addressing several common obstacles to renewable energy development and thereby facilitating its continued proliferation across the globe. In light of the invaluable role of innovation in the global energy transition, this final chapter offers an optimistic glimpse into the future by highlighting some specific ways that innovation could enable the renewable energy industry to grow and thrive for the balance of the twenty-first century.

Technological innovation

Technological advancements are the most familiar and well-publicized forms of energy-related innovation. Even relatively minor improvements in renewable energy system engineering and design can mitigate some of the tensions associated with renewable energy development.[1] Major advancements in renewable energy technology obviously have the potential to do much more, dramatically changing the siting and basic features of renewable energy projects in ways that eliminate many of the challenges outlined in this book. Of course, new technologies can also introduce new conflicts that necessitate their own unique policy solutions. The following are descriptions of three potentially game-changing renewable energy technologies that appear poised to play greater roles in the global energy economy over the next century—deep offshore wind energy, airborne wind energy, and energy storage systems—and discussions of the benefits and potential controversies that each could bring as it expands in the coming years.

Deep offshore wind energy

Many of the most impactful technological advancements for renewable energy over the next few decades are likely to involve "deep" offshore wind energy. The European Wind Energy Association defines deep offshore wind energy as the installation of wind turbines in waters more than 50 meters in depth.[2] Deep offshore wind energy technologies are nearly ready to hit the marketplace and have tremendous growth potential. Roughly 70 percent of the earth is covered with water,[3] and excellent wind energy resources are flowing above much of that vast area.[4] Deep offshore wind energy technologies have the potential to allow developers to site wind farms several miles further out at sea and access an enormous set of wind energy resources that are less vulnerable to many of the conflicts that complicate onshore wind farms.

Several characteristics of offshore wind energy make it a particularly attractive energy source. Winds blow about 40 percent more often offshore than onshore on average,[5] and there typically are no buildings, trees, or mountains capable of altering or diminishing wind currents over water.[6] Consequently, a wind turbine of a given size and type can be as much as 50 percent more productive if sited over the ocean than if it is installed at a good site on dry land.[7] Offshore turbines are also exposed to less turbulence because of the relatively steady flow of wind over water, enabling them to last longer and require less maintenance once installed.[8] And developers can erect much larger turbines over the sea than would ever be commercially viable onshore, where there are land-related constraints on turbine size.[9]

As offshore wind energy technologies improve, the relative costs of this type of development are decreasing and offshore wind is finally becoming commercially viable in more parts of the world. For instance, dramatic increases in offshore wind energy development are predicted for the North Sea in the next decade, with the United Kingdom planning to install 32 GW of offshore wind capacity in that sea between 2015 and 2025 alone.[10] After several years of lagging behind Europe, it appears that North America is finally poised to commence significant offshore wind energy development on that continent as well. A major offshore wind energy project is in the advanced planning stages off the shores of the Canadian province of British Columbia, and multiple offshore wind farms are purportedly in the works on both coasts of the United States.[11]

Even Japan, which continues to aggressively pursue alternative energy sources in the wake of the 2011 nuclear disaster at Fukishima,[12] is making unprecedented investments in offshore wind energy. With more total miles of coastline than the United States and one of the ten largest maritime "exclusive economic zones" in the world, Japan has tremendous offshore wind resources but has historically been limited in harnessing them because so few of its offshore areas have shallow water depths amenable to traditional anchored offshore turbines.[13] To overcome this water depth problem, the Japanese government has invested more than 22 billion yen ($226 million in U.S dollars) in the development of floating turbines near the Fukishima coastline with an intention to install a total of 140 floating turbines in the area capable of generating more than 1 GW of electricity by 2020.[14]

Deep offshore wind energy technologies like those being developed in Japan are becoming an increasingly attractive renewable energy strategy in countries throughout the world. Deep offshore wind energy is not just appealing because of the high quality of the wind resources involved; it is also far less likely to disrupt neighbors than onshore or shallow offshore wind energy development. Wind turbines installed a dozen miles offshore are practically invisible from coastlines and thus pose no risk of blemishing beachfront views. Their distance from homes and businesses also effectively eliminates any chance of noise disturbances, flicker effects, ice throws or most other sorts of conflict between wind turbines and neighbors.[15]

Deep offshore wind energy projects likewise avoid many of the transmission-related obstacles to onshore wind energy. As described in Chapter 7, much of the global population resides near coastlines, far from inland regions where the greatest concentrations of developable onshore wind are found. Because of this geographic mismatch between population centers and onshore wind resources, onshore wind energy development increasingly requires costly expansions of transmission infrastructure. Siting and funding these expansions can be prohibitively challenging, particularly when transmission routes must traverse jurisdictional boundaries. Such transmission constraints are potentially less of a concern in the context of deep offshore wind energy because deep offshore projects can be sited much closer to coastal cities. Far fewer miles of transmission infrastructure are required to deliver electricity from deep offshore wind farms to large populations of end users.

Despite the aforementioned advantages, deep offshore wind energy development obviously presents several unique challenges of its own. Perhaps most daunting among them are the logistical difficulties associated with installing gigantic wind turbines in unpredictable deep sea environments where waves, changing tides, high winds, and other uncontrollable factors can complicate construction.[16] Even after developers install them, offshore wind turbines and transmission lines themselves must be built to withstand storms, salt water corrosion, and the constant beating of ocean waves.[17] Technologies developed for offshore oil and gas extraction are somewhat helpful in the global effort to design deep offshore wind energy facilities, but the unique aspects of offshore wind energy projects still require their own distinct strategies and designs. Fortunately, costs associated with these new technologies are gradually coming down to the point that offshore wind energy is at last becoming more cost-competitive with other energy strategies.[18]

In addition to its logistical difficulties, deep offshore wind energy can also give rise to new types of conflicts with competing ocean uses such as shipping, fishing, and offshore oil and gas extraction. As the offshore wind energy industry expands, nations are increasingly adopting marine spatial planning programs to help balance these competing uses. Marine spatial planning programs effectively create an ocean zoning regime, using maps to limit particular types of marine uses to specifically identified areas.[19] Greater coordination among countries that share seas will also eventually be needed to help avoid disputes as deep offshore wind energy gradually increases the commercial value of marine areas in the coming years.[20]

Although deep offshore wind energy is presently little more than a fledgling niche industry, it unquestionably represents one of the next major horizons for the renewable energy movement. Given its myriad advantages, deep offshore wind energy development promises to be a vital component of the global energy transition over the next century as nations look for additional ways to add to their renewable energy portfolios without unduly disrupting other aspects of their economies.

Airborne wind turbines

Wind energy development is not only extending further and further offshore; it is also beginning to extend higher and higher into the airspace above land. The renewable energy potential of earth's high-altitude wind currents is astounding—a 2012 study estimated that the energy in these high winds is more than 100 times greater than the current global energy demand.[21] Practical constraints on turbine tower heights have historically kept these exceptional wind resources out of reach, but new airborne wind turbine technologies could soon make them readily accessible for electricity generation.

Airborne wind turbines are devices that soar thousands of feet above the surface of land, tethered to the ground by strong cables. These flying turbines convert the kinetic energy in high-altitude winds into electric current that is then delivered via the cables down to transmission facilities on the ground.[22] Federally-funded researchers in the United States have been exploring the possibilities of airborne wind energy for years and have compiled a fascinating array of futuristic turbine designs. Indeed, an airborne turbine can take the form of "a funnel-shaped blimp with a turbine at its back; or a balloon with vanes that rotate; a truss-braced wing; a parachute; a kite ... any and all of them are ideas being considered."[23] Multiple U.S. companies have already developed airborne turbine prototypes and have legitimate plans for commercial rollouts over the next several years.[24] In 2013, the possibilities of airborne wind energy proved promising enough to convince the Internet behemoth Google to purchase Makani Power—a company that is aggressively pursuing airborne wind technologies.[25]

Airborne wind energy innovations would not only provide access to a vast new renewable energy resource; they could also enable future developers to avoid some of the conflicts commonly associated with conventional wind farms. For example, many airborne wind turbines are designed to hover 1,000 feet or more above the ground[26]—heights at which neighbor disturbances such as turbine noise or flicker effects are less likely to occur. Because they would not be as permanently anchored to the ground as conventional turbines, airborne wind turbines could potentially be more easily relocated and could even be temporarily grounded during storms or particular periods when unblemished viewsheds were of greater priority. And because wind speeds tend to be stronger and more consistent at high altitudes, airborne wind technologies could allow for profitable wind energy development far closer to many metropolitan areas where low-altitude wind conditions are not favorable enough for land-based wind farms.[27]

Of course, airborne wind energy development would introduce its own distinct set of challenges as well. Airborne wind turbines would operate at altitudes rarely traversed by humans other than via an occasional helicopter or airplane. Because they would increase the productive value of high-altitude airspace, airborne wind energy technologies could generate new conflicts and legal questions surrounding the ownership and use of this space. In most areas within the United States, for instance, airspace more than 500 feet above the ground is classified as "navigable airspace"[28]—a commons or "public highway" for aviation.[29] Developers intending to operate airborne turbines in this high-altitude space would likely need exclusive rights to occupy it for that purpose and would want legal protection against disruptive air traffic in the area.

If airborne wind energy development eventually takes off and becomes a viable industry, courts and policymakers will be tasked with resolving the new legal issues it would raise. Would U.S. airborne wind farm developers need to lease navigable airspace from the government through a leasing system comparable to the one that presently exists for offshore wind farms and offshore oil drilling? Like deep offshore areas, navigable airspace is a sort of public trust resource that is largely controlled by the U.S. government, so some sort of federal leasing program for this space would seemingly be warranted.

In addition, airborne wind energy developers would need some limited property rights in surface areas directly below airspace leased for such projects so that they could tether the turbines to the ground and run transmission lines capable of delivering generated electricity to the grid. Except in cases where the airspace lease area is situated directly above federal public lands or above some other government property, these surface rights and the non-navigable airspace directly above them would be privately held and not government-controlled. An airborne wind farm developer would thus theoretically need to enter into two separate leases—one with the government and one with the private landowner—to install and operate an airborne turbine above any given parcel.

Of course, some would-be developers of airborne wind farms might wish to challenge the notion that federal leasing of navigable airspace for such projects is even necessary at all. An old U.S. Supreme Court case involving airspace declared that the owner of a parcel of land has private property rights in "at least as much of the space above the ground as he can occupy or use in connection with the land."[30] Prior to the advent of airborne wind energy technologies, no commercially viable, land-connected uses existed for high-altitude airspace so few challenged the federal government's control of it. Airborne wind turbines would change all of that, introducing a new, commercially valuable use for that airspace. There would seemingly still be good reason for the federal government to restrict and regulate uses of navigable airspace in spite of the advent of airborne wind energy technologies, but should the government be allowed to claim title to such airspace, lease it to private parties, and charge rent for its use? Arguably, restrictions on navigable airspace in a world of airborne wind turbines would most closely resemble zoning height restrictions that are commonly employed to limit uses of non-navigable space. Obviously, local governments do not typically assert property interests in such height-restricted airspace and attempt to lease it to underlying landowners. Indeed, an interesting debate involving developers, landowners, and governments could ultimately emerge over the balance between private and public property rights in navigable airspace in an era of airborne wind energy.

Eventually, airborne wind energy might also create a need for "spatial airspace planning" schemes similar to the marine spatial planning programs that increasingly govern ocean areas. Such zoning of navigable airspace would allow governments to designate and set aside certain airspace areas for airborne wind energy development while preserving other areas as air travel routes. This sort of airspace zoning could anger owners of lands situated directly below designated air travel routes because, at a minimum, it would effectively prohibit them from leasing their surface land and non-navigable airspace interests to airborne wind energy developers. Of course, the complex airspace ownership questions described in the previous paragraph would seem relevant in resolving the controversies resulting from these conflicts as well. In summary, airborne wind technologies could do much to facilitate continued growth in wind energy development, but could also give rise to some challenging new property and regulatory issues that would need to be resolved for such an industry to thrive.

Energy storage technologies

Innovations in energy storage also have the potential to serve critical functions in the global transition to sustainable energy. As mentioned in Chapter 7, wind and solar energy are, by their nature, intermittent energy resources—the wind is not always blowing and the sun is not always shining. Finding more efficient and cost-effective ways of storing wind- or solar-generated energy over time will thus be essential if these resources are ever to stand alone as electricity sources without the aid of backup power from fossil fuels or nuclear energy. Even when backup power sources are involved, low-cost energy storage technologies could greatly reduce the extent to which wind and solar energy resources complicate operators' efforts to balance power loads on the world's electricity grids.

Recognizing the importance of energy storage to the renewable energy movement, countries throughout the world are beginning to invest more attention and funding toward efforts to advance this growing research area. The German government plans to set aside roughly €200 million (US$270 million) toward developing better energy storage technologies, and the Japanese and United States governments have begun allocating more financial resources to this research as well.[31] In all, the research company IHS predicts that more than $19 billion will be invested worldwide in energy storage by the year 2017.[32]

The rechargeable lithium ion batteries that power common phones and flashlights have existed for decades but involve technologies that are far too expensive to employ on a large scale. Fortunately, numerous other energy storage strategies exist or are in development that could potentially prove far more cost-effective. Pumped storage facilities represent one interesting approach to storing energy that has existed for many years but has received increased attention as the wind and solar energy movements have grown. When there is an excess supply of electricity on the grid, a pumped storage facility uses that energy to pump water from one reservoir to another reservoir located uphill. Then, when electricity demand on the grid rises, facility operators release water from the upper reservoir, sending it through hydro-turbines that generate electricity as the water makes its way down to the lower reservoir.[33] Eventually, when there is once again an excess supply of electricity on the grid, the facility uses this low-cost electricity to pump water back into the upper reservoir so that the cycle can continue.

One major disadvantage of pumped storage facilities is that they are a relatively inefficient means of storing energy. The developer of one modern pumped storage facility noted that the system was capable of providing only about 3 kWh of future electricity for every 4 kWh it consumed—an energy loss of roughly 33 percent.[34] For obvious reasons, pumped storage facilities are also rarely cost-justifiable in locations that lack natural elevation changes or an ample supply of available water. Nonetheless, pumped storage systems can sometimes be a useful way of mitigating the intermittency problems associated with renewables. They tend to be most useful when sited near large solar energy projects or wind farms, where they can make use of the transmission infrastructure already in place for those projects.

A handful of other energy storage strategies have likewise been around for several years but have thus far proven to be limited in their usefulness. Compressed air systems, which have existed for decades but are relatively few in number, fall into this category of storage strategies. Compressed air systems involve the use of excess electricity to compress and pump air into sealed underground reservoirs. These systems can then produce electricity at a later time by slowly releasing the compressed air to rotate turbines near the surface.[35] Like pumped storage facilities, compressed air systems are relatively energy-inefficient and are usable only in areas with suitable geographic features. These and other limitations prevent them from contributing in a major way to energy storage challenges in many regions.

Spinning flywheel technologies, which use devices similar to a spinning top toy to store energy, have also existed for many years. Because flywheels are subject to fewer geographic constraints than pumped storage facilities or compressed air systems and are usable on a much smaller scale, they have shown some promise as a potentially valuable energy storage strategy in certain contexts. However, because they are presently capable of storing energy for only relatively short periods of time, flywheels are unlikely to ever become a dominant means of energy storage.[36]

Fortunately, new energy storage technologies are beginning to emerge that provide reasons to hope for real progress on this front. For instance, a molten salt energy storage system is helping to store energy generated at the Solana project—a large concentrating solar plant near Gila Bend, Arizona, in the United States.[37] Numerous carefully positioned mirrors at the plant reflect sunlight to heat large tanks of molten salt. These molten salt tanks are specially designed to retain much of their heat, enabling them to continue producing steam-generated electricity for several hours after the sun has set. Although these innovations are only available in certain settings, they have the potential to reduce the intermittency of power supplies at some utility-scale solar energy projects.

"Smart" battery system technologies are another bright spot in the energy storage realm. These systems seek to smooth out the intermittency of distributed solar energy systems or to reduce the variability of an electricity customer's demand. Researchers are increasingly exploring ways to better integrate energy storage systems with solar arrays so that the arrays can feed a more consistent supply of electricity onto the grid.[38] Similarly, companies are beginning to market distributed energy storage systems to businesses operating in areas where electricity prices fluctuate throughout the day, surging to expensive levels during periods of peak demand. Energy storage systems can allow businesses in these areas to purchase less power during peak use periods and thereby reduce their electricity bills.[39]

Utility mandates for energy storage—policies patterned after renewable portfolio standards that effectively require utilities to invest in energy storage systems—have the potential to be a valuable tool for promoting advancements in energy storage technologies. In late 2013, California became the first state in the United States to mandate that its major power companies purchase energy storage capacity to help mitigate grid-related impacts from the state's growing use of intermittent energy sources such as wind and solar energy.[40] This mandate issued by the California Public Utility Commission will not only help facilitate continued renewable energy development in the state; it will also spur additional investment in mid- to large-scale energy storage research and thereby help to accelerate innovation in this important industry. Affordable energy storage technologies of all sizes will be needed as the world inches ever closer toward the utopian goal of a truly sustainable energy system.

Policy innovation

Although advancements in renewable energy *policy* tend to garner far less attention than innovations in *technology*, policy ingenuity also has a vital role to play in furthering the global transition to sustainable energy. Feed-in tariffs, mandatory net metering programs, tax credit schemes, renewable portfolio standards, solar access, solar rights statutes, and dozens of other creative policies have already done much to propel renewable energy growth throughout the world. However, significantly more policy innovation will be needed for the renewable energy movement to steadily progress throughout the twenty-first century.

Renewable energy technologies have developed at a blistering pace in recent years, achieving ever greater levels of efficiency and sophistication. As highlighted in this book, these advancements are raising numerous legal and policy questions that remain unaddressed in many jurisdictions around the world. In this environment of rapid innovation, stale policies could unnecessarily hinder the global energy transition unless policymakers manage to keep up with all of these technological changes.

Among other things, policymakers are increasingly grappling with how to clarify and allocate property interests in some natural resources that had little commercial value until the advent of renewable energy technologies. For instance, in this era of solar panels, should policymakers reconsider laws governing landowners' rights to have direct sunlight strike their land? Given the growing value of wind resources, should landowners possess any legally protected property rights in the wind that passes over their parcels? And how, if at all, should property law respond to the the advent of deep offshore and airborne wind technologies? In jurisdictions where these sorts of questions are left unaddressed, uncertainty about the nature and scope of the property rights involved in renewable energy development can chill investment and hinder valuable projects. As suggested in earlier chapters, clear, carefully-tailored property rights regimes are paramount in the effort to overcome uncertainty in these contexts.

Innovative policies are also needed to govern the ever growing list of tensions between renewable energy development and traditional land uses. What laws are best equipped to balance the goal of promoting renewable energy with other, competing policy goals such as historic preservation, wildlife conservation, homeland security, agriculture, and recreation? Because conflicts with each of these categories of land uses involve their own unique challenges and issues, policymakers will need to be more inventive than ever in structuring policy solutions to address them. The global effort is well underway to craft policies that reconcile the push for renewable energy with other policy goals, but it is bound to require far more attention from academics, legislators, and others in the coming years.

In addition to effectively addressing new legal questions and tensions raised by renewable energy, policymakers can also assist the global energy transition by crafting low-cost policies to help drive additional demand for renewables and ease the development process. Recently, as budgetary and political factors in the United States have caused federal- and state-level support for renewables to wane, local governments have stepped in and pioneered several inventive new policy strategies of their own. If financial subsidies for renewables continue to dry up over time and utilities increasingly view renewables as a competitive threat, this grassroots policymaking in cities and towns will be ever more important.

"Solar mandate" policies are one example of a local policy innovation capable of significantly increasing distributed solar energy installations at a relatively low government cost. Solar mandates are local building or zoning code provisions that require citizens to install on-site solar energy systems or fund off-site solar energy projects as a condition for receiving real estate development approvals. Municipalities can impose these mandates on almost any type of real estate development, from residential subdivisions to condominiums to commercial and industrial projects. The mandates can also be incorporated into building codes and enforced in connection with building permit applications.

Solar mandates offer significant advantages over some other common policy approaches to promoting solar energy at the local level. For example, in jurisdictions with established permitting and approval processes for real estate development, local governments can cheaply and easily adopt and implement solar mandate policies by merely incorporating the mandates into their existing sets of development conditions. By increasing the cost of new real estate development within a jurisdiction, solar mandate requirements can also bolster real estate values in the communities that adopt them. Because of these and other potential benefits, more and more cities and counties in the United States are beginning to adopt solar mandates and similar laws. The city of Lancaster, California, adopted a very aggressive solar mandate policy in March of 2013, requiring solar PV installations with minimum nameplate capacities of between 0.5kW and 1.5kW on every new housing unit within the city.[41] The California town of Sebastopol adopted its own solar mandate policy only a couple of months later.[42] And municipalities in the U.S. states of Colorado and Idaho have had related laws on the books for several years.[43] As direct government subsidies for solar energy fade away, these sorts of mandates could be a powerful and affordable means for local governments to help perpetuate solar energy growth.

Other local policies, like the Klickitat County, Washington, energy overlay zoning scheme described in the previous chapter, can facilitate renewable energy development in locales where it is most welcome by streamlining permit approvals and lowering regulatory hurdles. By sending clear signals to developers and reducing the regulatory and transaction costs, these low-cost policies can do much to accelerate the pace of wind and solar energy installations in jurisdictions that adopt them.

In addition to the few examples just mentioned, there are undoubtedly countless other policy improvements waiting to be discovered that are capable of lending invaluable help to the global transition toward more sustainable energy practices. In light of these opportunities, legal academics and policymakers in the area of renewable energy should endeavor to be as innovative as the engineers and developers, thinking creatively and continually searching for ways to improve the legal structures that govern this important industry.

A bright future

This book has provided just one snapshot of the still-fledgling renewable energy movement—a movement that is advancing at a frenetic pace but still faces countless uncertainties and challenges. Within a few short years, much of this book will surely become outmoded as emerging technologies continue to reshape the renewable energy industry and raise ever newer legal questions and logistical problems. Meanwhile, there remains an endless amount of work to be done to ensure that the planet makes full productive use of the impressive clean energy technologies that are already at its disposal. The earth has only a finite supply of non-renewable energy reserves, and the realities of climate change are becoming more and more apparent every year. In the face of these difficulties, humankind's long-term well-being may hinge on whether we are diligent and persistent enough to achieve the idyllic goal of true global energy sustainability.

Stalled progress in international climate change talks and the unrelenting construction of coal-fired power plants in China and other developing countries might justifiably cause some to doubt that worldwide sustainable energy is even possible. Without question, more unforeseen hurdles will get in the way of renewable energy development as it continues to spread across the planet. Recent incidents involving the criminal dismantling of wind turbines in the Philippines for their copper and other valuable materials exemplify the sorts of unpredictable obstacles that could yet emerge as wind and solar energy become ever more commonplace across the globe.[44]

Nonetheless, there is great reason for optimism. Increasingly, individuals, companies, cities, and nations are installing renewable energy systems because they have concluded that doing so is truly in their own best economic interests. Falling solar PV costs are leading more and more homeowners to install solar energy systems on their homes as a way to lock in and reduce their electricity expenses. As of 2013, the retail giant Wal-Mart owned more solar energy generating capacity than the total amount installed in 38 U.S. states combined.[45] And the oil-rich country of Saudi Arabia is investing roughly $109 billion in pursuit of its ambitious goal of meeting one third of its electricity needs through solar power by the year 2032.[46]

These trends suggest that simple economics and market incentives might eventually become the single greatest driver of the global transition to energy sustainability—not United Nations protocols, carbon taxes, cap-and-trade schemes, or other policy tools. No other scenario would be more ideal for the renewable energy sector because genuine market incentives tend to be far more effective and efficient at spurring concerted worldwide action than any regulation or tax program. In the meantime, the work of technological and policy innovation must march aggressively forward, surmounting barriers one by one until the sustainable energy movement becomes fully equipped to sustain itself.

Notes

1 For instance, scientists in Germany have developed ways to reduce the loudness of wind turbines that could help to reduce conflicts over turbine noise. *See* Eoin O'Carroll, *German Scientists Develop Silencer for Wind Turbines*, CHRISTIAN SCIENCE MONITOR (Aug. 22, 2008), available at www.csmonitor. com/Environment/Bright-Green/2008/0822/scientists-develop-silencer-for-wind-turbines (last visited Nov. 7, 2013).

2 *See* Athanasia Arapogianni & Anne-Benedicte Genachte, *Deep Water: The Next Step for Offshore Wind Energy* at 7, European Wind Energy Association (July 2013), available at www.ewea.org/fileadmin/files/library/publications/reports/ Deep_Water.pdf (last visited July 17, 2013).

3 *See* id.

4 For example, offshore wind resources in the first 50 nautical miles off of the U.S. coast total more than 4,000 GW—about four times the nation's current electricity needs. *See* Walter Musial & Bonnie Ram, *Large-Scale Offshore Wind Power in the United States* at 4, NREL/TP-500-49229 (Sept. 2010).

5 *See* Daina Lawrence, *Canada, U.S. Poised to Catch the Offshore Wind*, GLOBE AND MAIL, 2013 WLNR 26579699 (Oct. 23, 2013).

6 *See* Todd J. Griset, *Harnessing the Ocean's Power: Opportunities in Renewable Ocean Energy Resources*, 16 OCEAN & COASTAL L.J. 395, 398–99 (2011).

7 *See* Margaret Bryant, *Wind Energy in Texas: An Argument for Developing Offshore Wind Farms*, 4 ENVT'L. & ENERGY L. & POL'Y J. 127, 131–32 (2009).

8 *See* Carolyn S. Kaplan, Esq., *Congress, the Courts, and the Army Corps: Siting the First Offshore Wind Farm in the United States*, 31 B.C. ENVTL. AFF. L. REV. 177, 191 (2004).

9 Musial & Ram, *supra* note 4 at 4.

10 *See* Climate Wire, *Offshore Wind Turbines Keep Growing in Size*, SCIENTIFIC

AMERICAN (Sept. 19, 2011), available at www.scientificamerican.com/article. cfm?id=offshore-wind-turbines-keep (last visited Oct. 11, 2013).

11 *See* Lawrence, *supra* note 5.

12 *See* Carol J. Williams, *Japan Gives Wind Energy a Spin*, LOS ANGELES TIMES (Nov. 11, 2013) (describing the Japanese citizenry's continued opposition to nuclear energy and growing investments in renewables).

13 *See* Hiroko Tabuchi, *To Expand Offshore Power, Japan Builds Floating Windmills*, NEW YORK TIMES (Oct. 24, 2013).

14 *See* id.

15 Detailed descriptions of these sorts of conflicts are set forth in Chapter 2.

16 For a description of some of these challenges, *see generally* Edward Platt, *The London Array: The World's Largest Offshore Wind Farm*, THE TELEGRAPH (July 28, 2012), available at www.telegraph.co.uk/earth/energy/windpower/9427156/ The-London-Array-the-worlds-largest-offshore-wind-farm.html (last visited Oct. 15, 2013).

17 *See* Musial & Ram, *supra* note 4 at 4.

18 *See* Lawrence, *supra* note 11 (quoting an industry official as stating that the "price gap between clean energy and dirty energy has narrowed, the technology has advanced so much and the politics have changed" in favor of offshore wind energy development).

19 *See* Michelle E. Portman, et al., *Offshore Wind Energy Development in the Exclusive Economic Zone: Legal and Policy Supports and Impediments in Germany and the US*, 37 ENERGY POLICY 3596, 3604 (2009).

20 *See* Kara M. Blake, *Marine Spatial Planning for Offshore Wind Energy Projects in the North Sea: Lessons for the United States* at 54, Master's thesis (University of Washington) (2013), available at https://digital.lib.washington.edu/research-works/bitstream/handle/1773/22801/Blake_washington_0250O_11501. pdf?sequence=1 (last visited Oct. 17, 2013); Julia Lisztwan, *Stability of Maritime Boundary Agreements*, 37 YALE J. INT'L L. 153, 154 (2012).

21 Brian Clark Howard, *Turbines Ready for Takeoff*, nationalgeographic.com (Sept. 24, 2012), available at http://news.nationalgeographic.com/news/ energy/2012/09/pictures/120924-flying-wind-turbines/ (last visited Oct. 25, 2013) (citing Kate Marvel, et al., *Geophysical Limits to Global Wind Power*, 3 NATURE CLIMATE CHANGE 118 (2012).

22 *See* Jim Hodges, *An Answer to Green Energy Could Be in the Air*, NAT'L AERONAUTICS & SPACE ADMIN. (Dec. 10, 2010), www.nasa.gov/topics/ technology/features/capturingwind.html (last visited Oct. 25, 2013).

23 Id.

24 For more information about some of these companies, *see, e.g.*, www.altaerosenergies .com/&http://www.makanipower.com/home/ (last visited Oct. 25, 2013).

25 *See* Frederic Lardinois, *Google X Acquires Makani Power and Its Airborne Wind Turbines*, techcrunch.com (May 22, 2013), available at http://techcrunch. com/2013/05/22/google-x-acquires-makani-power-and-its-airborne-wind-tur- bines/ (last visited Oct. 25, 2013).

26 *See* Dave Levitan, *High-Altitude Wind Energy: Huge Potential—And Hurdles*, YALE ENVIRONMENT 360 (Sept. 24, 2012), available at http://e360.yale.edu/ feature/high_altitude_wind_energy_huge_potential_and_hurdles/2576/　(last visited Oct. 25, 2013).

27 *See* Brian MacCleery, *The Advent of Airborne Wind Power*, WIND SYSTEMS (January 2011), available at http://windsystemsmag.com/article/detail/187/ the-advent-of-airborne-wind-power (last visited Oct. 25, 2013).

28 *See* 14 C.F.R. § 77.23 (2009).

29 *See United States v. Causby*, 328 U.S. 256, 264 (1946).

30 Id. at 264 (citing *Hinman v. Pac. Air Lines Trans. Corp.*, 84 F.2d 755 (9th Cir. 1936)).

31 *See* Kate Galbraith, *Filling the Gaps in the Flow of Renewable Energy*, NEW YORK TIMES (Oct. 3, 2013), available at www.nytimes.com/2013/10/03/business/energy-environment/Filling-the-Gaps-in-the-Flow-of-Renewable-Energy.html?_r=0 (last visited Oct. 9, 2013).

32 *See* Peter Kelly-Detwiler, *Energy Storage: Continuing to Evolve*, Forbes.com (May 15, 2013), available at www.forbes.com/sites/peterdetwiler/2013/05/15/energy-storage-continuing-to-evolve/ (last visited Oct. 25, 2013).

33 *See* David Lindley, *The Energy Storage Problem*, 463 NATURE 18 (2010).

34 *See* Tildy Bayar, *The Time Is Right for Small Pumped Energy Storage, UK Hydropower Developer Says*, renewableenergyworld.com (Sept. 10, 2013), available at www.renewableenergyworld.com/rea/news/article/2013/09/the-time-is-right-for-small-pumped-storage-in-the-uk-developer-says (last visited Oct. 26, 2013).

35 *See* Lindley, *supra* note 33 at 19.

36 *See* id. at 20.

37 *See* Matthew L. Wald, *Arizona Utility Tries Storing Solar Energy for Use in the Dark*, NEW YORK TIMES (Oct. 18, 2013).

38 *See* Kelly-Detwiler, *supra* note 32.

39 *See* Ucilia Wang, *How $5M Will Help a Startup Tackle the Emerging Energy Storage Market*, FORBES (Oct. 24, 2013), available at www.forbes.com/sites/uciliawang/2013/10/24/how-5m-will-help-a-startup-tackle-the-energy-storage-market/ (last visited Oct. 30, 2013).

40 *See* Dana Hull, *California Adopts First-in-nations Energy Storage Plan*, SAN JOSE MERCURY NEWS (Oct. 17, 2013), available at www.mercurynews.com/business/ci_24331470/california-adopts-first-nation-energy-storage-plan (last visited Oct. 30, 2013).

41 *See* Lancaster Municipal Code §§ 17.08.060 and 17.08.305 (2013).

42 *See* Sebastapol Municipal Code § 15.72.040 (2013).

43 *See generally* Boulder County BuildSmart Code § N1105.2.5.4 (2012) Tables 3 and 4 of the Boulder County BuildSmart Code, IRC Chapter 11, available at www.bouldercounty.org/doc/landuse/2012buildsmartcode.pdf.; Pitkin County Code § 11.32, Appendix A & B (2000); Blaine County Code § 7-6-6, Appendix A (2011).

44 *See* Dermot Duncan & Benjamin K. Sovacool, *The Barriers to the Successful Development of Commercial Grid Connected Renewable Electricity Projects in Australia, Southeast Asia, the United Kingdom and the United States of America*, 2 RENEWABLE ENERGY L. & POL'Y REV. 283, 292 (2011).

45 *See* Tom Randall, *Wal-Mart Now Draws More Solar Power Than 38 U.S. States*, Bloomberg.com (Oct. 25, 2013), available at www.bloomberg.com/news/2013-10-24/wal-mart-now-has-more-solar-than-38-u-s-states-drink-.html (last visited Nov. 5, 2013).

46 *See* Wael Mahdi & Marc Roca, *Saudi Arabia Plans $109 Billion Boost for Solar Power*, Bloomberg.com (May 12, 2012), available at www.bloomberg.com/news/2012-05-10/saudi-arabia-plans-109-billion-boost-for-solar-power.html (last visited Nov. 5, 2013).

Bibliography and sources

Acher, J. (2011) "Next Thing in Wind Energy: Stealth Turbines" [online] 29 June. Available at: www.reuters.com/article/2011/06/29/us-windenergy-stealth-vestas-idUSTRE75S2JM20110629 [Accessed: 22 June 2012].

Agnew, D. (2013) "Responsibility to Future Generations: Renewable Energy Development on Tribal Lands" *Whitehouse.gov.* [online] Available at: www.whitehouse.gov/blog/2013/06/25/responsibility-future-generations-renewable-energy-development-tribal-lands [Accessed: 12 August 2013].

Alpert, D. (2012) "Preservation Staff Reject Solar Panels on Cleveland Park Home" [online] 31 May. Available at: http://greatergreaterwashington.org/post/15012/preservation-staff-reject-solar-panels-on-cleveland-park-home [Accessed: 19 August 2013].

Altaeros Energies. (n.d.) "Altaeros Main Page" [online] Available at: www.altaeros energies.com [Accessed: 25 October 2013].

Alvarez, T. L. (2007) "Don't Take My Sunshine Away: Right-to-Light and Solar Energy in the Twenty-First Century" *Pace L. Rev.*, 28 p. 535.

American Wind Energy Association. (2008) *Wind Energy Siting Handbook 5.4.1.* Available at: www.awea.org/sitinghandbook/downloads/Chapter_5_Impact_Analysis_and_Mitigation.pdf [Accessed: 21 June 2012].

Anders, S., Grigsby, K., and Kurkuk, C. (2007) "California's Solar Shade Control Act: A Review of the Statutes and Relevant Cases" *Energy Policy Initiatives Center, University of San Diego School of Law.* Available at: www.sandiego.edu/documents/epic/100329_SSCA_Final_000.pdf [Accessed: 22 February 2014].

Arapogianni, A. and Genachte, A. (2013) "Deep Water: The Next Step for Offshore Wind Energy" *European Wind Energy Association.* [online] Available at: www.ewea.org/fileadmin/files/library/publications/reports/Deep_Water.pdf [Accessed: 17 July 2013].

Arnett, E. B., Huso, M. M., Schirmacher, M. R., and Hayes, J. P. (2010) "Altering Turbine Speed Reduces Bat Mortality at Wind-Energy Facilities" *Frontiers in Ecology and the Environment*, 9(4), p. 209.

Azau, S. (2011) "Nurturing Public Acceptance" *Wind Directions*, Iss. 30, pp. 30–36.

Baerwald, E. F., D'Amours, G. H., Klug, B., and Barclay, R. M. (2008) "Barotrauma Is a Significant Cause of Bat Fatalities at Wind Turbines" *Current Biology*, 18(16), pp. 695–96.

Bakewell, S. (2013) "U.K. to Share More Wind-Power Wealth With Local Residents" *Bloomberg.* [online] Available at: www.bloomberg.com/news/2013-06-05/uk-to-share-more-wind-power-wealth-with-local-residents.html [Accessed: 6 June 2013].

Bald and Golden Eagle Protection Act of 1940. (1972) United States Code Annotated, Title 16, Chapter 5A, Subchapter II, Section 668(a); 50 Code of Federal Regulations 22.3. Taking and using the bald and golden eagle for scientific, exhibition, and religious purpose.

Barbose, G. (2012) "Tracking the Sun V: An Historical Summary of the Installed Price of Photovoltaics in the United States from 1998 to 2011' *Emp.lbl.gov.* [online] Available at: http://emp.lbl.gov/publications/tracking-sun-v-historical-summary-installed-price-photovoltaics-united-states-1998-2011 [Accessed: 14 June 2013].

Barringer, F. (2008) "Trees Block Solar Panels, and a Feud Ends in Court" [online] 7 April. Available at: www.nytimes.com/2008/04/07/science/earth/07redwood.html?pagewanted=all&_r=0 [Accessed: 28 June 2013].

Barry, S. (2009) "Indigenous Backing for Cooktown Wind Project" [online] 11 August. Available at: www.abc.net.au/news/2009-08-12/indigenous-backing-for-cooktown-wind-project/1387704 [Accessed: 12 August 2013].

Bayar, T. (2013) "The Time Is Right for Small Pumped Energy Storage, UK Hydropower Developer Says" *Renewable Energy World.* [online] Available at: www.renewableenergyworld.com/rea/news/article/2013/09/the-time-is-right-for-small-pumped-storage-in-the-uk-developer-says [Accessed: 26 October 2013].

Beck, H. (2011) "Cooktown One Step Closer to Wind Farm" [online] 12 May. Available at: www.cairns.com.au/article/2011/05/12/163591_local-news.html [Accessed: 12 August 2013].

Behles, D. (2012) "An Integrated Green Urban Electrical Grid" *Wm. & Mary Envtl. L. & Pol'y Rev.*, 36(3), pp. 677–78.

Beitsch, R. (2011) "Bill Requires Wind Turbine Compensation Sharing" [online] 7 March. Available at: http://bismarcktribune.com/news/local/govt-and-politics/2011_session/bill-requires-wind-turbine-compensation-sharing/article_f3a7baaa-3277-11e0-82c9-001cc4c002e0.html [Accessed: 24 June 2012].

Benfield, K. (2012) "Can Solar Panels and Historic Preservation Get Along?" *The Atlantic Cities.* [online] Available at: www.theatlanticcities.com/design/2012/06/can-solar-panels-and-historic-preservation-get-along/2364/ [Accessed: 19 August 2013].

Benson, E. D., Hansen, J. L., Schwartz Jr., A. L., and Smersh, G. T. (1998) "Pricing Residential Amenities: The Value of a View" *The J. of Real Estate Fin. & Econ.*, 16(1), p. 55.

Bertsch, D. (2011) "When Good Intentions Collide: Seeking a Solution to Disputes between Alternative Energy Development and the Endangered Species Act" *Sustainable Dev. L.J.*, 14, p. 74.

Black's Law Dictionary. (1968) 4th ed. St. Paul, MN: West Publishing.

Black's Law Dictionary. (2004) 8th ed. Boston: West Group.

Blake, K. M. (2013) *Marine Spatial Planning for Offshore Wind Energy Projects in the North Sea: Lessons for the United States.* Master's thesis. University of Washington. Available at: https://digital.lib.washington.edu/researchworks/bitstream/handle/1773/22801/Blake_washington_0250O_11501.pdf?sequence=1 [Accessed: 17 October 2013].

Blum, J. (2005) "Researchers Alarmed by Bat Deaths from Wind Turbines" [online] 1 January. Available at: www.washingtonpost.com/wp-dyn/articles/A39941-2004Dec31.html [Accessed: 12 June 2013].

Booker, C. (2010) "Wind Turbines: Eco-friendly—But Not to Eagles" [online] 13 March. Available at: www.telegraph.co.uk/comment/columnists/

christopherbooker/7437040/Eco-friendly-but-not-to-eagles.html   [Accessed: 12 June 2013].

Boubekri, M. (2005) "Solar Access Legislation—A Historical Perspective of New York City and Tokyo" *Planning & Envtl. L.*, 57(5), p. 3.

Bowater, D. and Mason, R. (2011) "Blown Away: Gales Wreck Wind Turbines and Scottish Storm Wreaks Havoc" *The Telegraph*, 9 December.

Boyles, J. G., Cryan, P. M., McCracken, G. F., and Kunz, T. H. (2011) "Economic Importance of Bats in Agriculture" *Science*, 332(6025), pp. 41–42.

Brenner, M. (2008) *Wind Farms and Radar*. [report] McLean: Federation of American Scientists, pp. 8–9. Available through: www.fas.org/irp/agency/dod/jason/wind.pdf [Accessed: 1 March 2011].

Bridgeman, S. (2012) "Who Owns the Wind?" [online] 13 September 2012. Available at: www.nzherald.co.nz/lifestyle/news/article.cfm?c_id=6&objectid=10833580 [Accessed: 16 August 2013].

Brisman, A. (2005) "The Aesthetics of Wind Energy Systems" *NYU Envtl. L.J.*, 13, pp. 8–46.

Bronin, S. C. (2010) "Curbing Energy Sprawl with Microgrids" *Conn. L. Rev.*, 43, p. 547.

Bronin, S. C. (2012) "The Promise and Perils of Renewable Energy on Tribal Lands" *HeinOnline*, p. 221. http://home.heinonline.org

Brown, A. C. and Rossi, J. (2010) "Siting Transmission Lines in a Changed Milieu: Evolving Notions of the Public Interest in Balancing State and Regional Considerations" *U. Colo. L. Rev.*, 81, p. 717.

Bryant, M. (2009) "Wind Energy in Texas: An Argument for Developing Offshore Wind Farms" *Envtl. & Energy L. & Pol'y J.*, 4, pp. 131–32.

Bryce, R. (2010) "The Brewing Tempest over Wind Power" [online] 1 March. Available at: http://online.wsj.com/article/SB1000142405274870424000457508 5631551312608.html [Accessed: 19 June 2012].

Burch v. NedPower Mount Storm, LLC, 647 S.E.2d 879, 893–94 (W. Va. Ct. App. 2007).

Bureau of Land Management. (2014) *Right-of-Way Management, Solar and Wind Energy*. Instruction Memorandum No. 2011-061. [report] Washington, D.C.: United States Department of Interior.

Burleson, E. (2009) "Wind Power, National Security, and Sound Energy Policy" *Penn State Envtl. L., Rev*, 17, p. 137.

Butler, S. H. (2009) "Headwinds to a Clean Energy Future: Nuisance Suits Against Wind Energy Projects in the United States" *California L. Rev.*, pp. 1337–69.

California Civil Code 714. (1978) Section 714, Division 2, Part 1, Title 2, Chapter 2, Article 2. Property; ownership; modifications of ownership; conditions of ownership; reasonable restrictions on solar energy systems.

California Public Resources Code. (2008) Chapter 12, Sections 25980–25986. Solar shade control.

California Revenue and Tax Code (1976). Section 23601. Solar energy tax credit as an incentive for taxpayers to install "solar energy systems."

California Wilderness Coalition v. Dep't of Energy, 631 F.3d 1072 (9th Cir. 2011).

Cameron, W. (2005) "Chinooks—Warm West Winds" *MountainNature.com*. [online] Available at: www.mountainnature.com/climate/Chinook.htm [Accessed: 19 November 2013].

Campoy, A. and Simon, S. (2011) "Wind Fuels Fight in Oil Patch" *Wall Street Journal*, 26 November.

Cannon Power Group. (2014) "Windy Point/Windy Flats, State of Washington" [online] Available at: www.cannonpowergroup.com/wind/projects/wp-wf/ [Accessed: 10 December 2013].

Cardwell, D. (2013) "Intermittent Nature of Green Power Is Challenge for Utilities" *New York Times*, 15 August, p. B1.

Casper Logistics Hub. (n.d.) "Top 20 States with Wind Energy Potential" [online] Available at: www.casperlogisticshub.com/downloads/Top_20_States.pdf [Accessed: 3 September 2013].

Clarke, C. (2013) "California Governor Signs Controversial Solar Bill, Among Others" *KCET*. [online] Available at: www.kcet.org/news/rewire/government/ governor-signs-controversial-solar-bill.html [Accessed: 8 October 2013].

Cleveland Park Historical Society. (n.d.) "Doing Work on Your Historic District Home" [online] Available at: www.clevelandparkhistoricalsociety.org/historic- district/doing-work-on-your-historic-district-home [Accessed: 19 August 2013].

Climate Wire. (2011) "Offshore Wind Turbines Keep Growing in Size" [online] 19 September. Available at: www.scientificamerican.com/article.cfm?id=offshore- wind-turbines-keep [Accessed: 11 October 2013].

Colette, M. (2012) "Kingsville Officials Opposed Wind Farm Near Riviera: Navy Says Impact Will Be Minimal" [online] 28 March. Available at: www.caller.com/ news/2012/mar/28/kingsville-officials-oppose-wind-farm-near-navy [Accessed: 22 June 2012].

Colorado Revised Statutes. Title 38, Article 32.5, Section 100. Residential sectors; Applicable technologies: passive solar, active space and water heating.

Conn. Gen. Stat. Ann. (1957) Chapter 7, Section 147. Municipalities; regulation of obstructions in waterways.

Contra Costa Water Dist. v. Vaquero Farms, Inc., 58 Cal. App. 4th 883 (Cal. Ct. App. 1997).

Copping, J. (2011) "Switch-Off for Noisy Wind Farms" [online] 19 November. Available at: www.telegraph.co.uk/earth/energy/windpower/8901431/Switch-off- for-noisy-wind-farms.html [Accessed: 19 June 2012].

Cryan, P. M. (2011) "Wind Turbines as Landscape Impediments to the Migratory Connectivity of Bats." *Envtl. L.(00462276)*, 41(2), pp. 357–67.

Cummings, J. (2012) "Wind Farm Noise and Health: Lay Summary of New Research Released in 2011" [online] April. Available at: www.acousticecology. org/wind/winddocs/AEI_WindFarmsHealthResearch2011.pdf [Accessed: 20 June 2012].

Dahl, E. L., May, R., Hoel, P. L., Bevanger, K., Pedersen, H. C., Roskaft, E., and Stokke, B. G. (2013) "White-tailed Eagles (Haliaeetus Albicilla) at the Smola Wind-power Plant, Central Norway, Lack Behavioral Flight Responses to Wind Turbines" *Wildlife Society Bulletin*, 37(1), pp. 66–74.

Danelski, D. (2012) "Mojave Desert: Military Wants to Limit Wind Development" [online] 20 May. Available at: www.pe.com/local-news/topics/topics-environment- headlines/20120520-mojave-desert-military-wants-to-limit-wind-development. ece [Accessed: 22 June 2012].

Darian-Smith, E. (2010) "Environmental Law and Native American Law" *Ann. Rev. of L. & Soc. Sci.*, 6, p. 361.

Davenport, P. (2007) "Arizona Regulators Skeptical of New Electric Line to California" *U-T San Diego*. [online] Available at: http://legacy.utsandiego.com/ news/business/20070530-1352-wst-sharingpower.html [Accessed: 19 September 2013].

Davidson, R. (2013) "Grid Must Catch Up With Renewables Growth" *Windpower Monthly*. [online] Available at: www.windpowermonthly.com/article/1209701/ grid-catch-renewables-growth [Accessed: 24 September 2013].

De La Torre, K. and Thompson III, R. S. (2011)*The Indian Energy Promotion and Parity Act of 2010: Opportunities for Renewable Energy Projects in Indian Country*. 2011 No. 2 RMMLF-INST Paper No. 8. [report] Rocky Mountain Mineral Law Foundation.

Deline, C., Meydbray, J., Donovan, M., and Forrest, J. (2012) *Photovoltaic Shading Testbed for Module-level Power Electronics*. TP-5200-54876. [report] Golden, Co: NREL.

D'Emilio, F. (2007) "Vatican Pushes Solar Energy" *Boston Globe*, 6 June.

DePaolo, D. (2013) "Oglala Sioux Make Strides Toward Energy Independence" [online] 15 July. Available at: www.ksfy.com/story/22847509/oglala-sioux-make-strides-toward-energy-independence [Accessed: 12 August 2013].

Desholm, M. (2009) "Avian Sensitivity to Mortality: Prioritising Migratory Bird Species for Assessment at Proposed Wind Farms" *J. of Envtl. Manage.*, 90(8), p. 2672.

Diamond, K. E. and Crivella, E. J. (2011) "Wind Turbine Wakes, Wake Effect Impacts, and Wind Leases: Using Solar Access Laws as the Model for Capitalizing on Wind Rights During the Evolution of Wind Policy Standards" *Duke Envtl. L. & Pol'y F.*, 22, p. 195.

Dickerson, M. (2008) "Hey, your shade trees are blocking my solar panels" *The Los Angeles Times*, 15 November.

Dimugno, L. (2012) "Another Wind Turbine Blaze: Fire Breaks Out at Iowa Wind Farm" [online] 24 May. Available at: http://nawindpower.com/e107_plugins/ content/content.php?content.9883 [Accessed: 25 June 2012].

Donovan, L. (2008) "Two Energy Products Competing for the Wind" [online] 22 February. Available at: http://bismarcktribune.com/news/local/ two-energy-projects-competing-for-the-wind/article_4bd1f0d6-6616-512b-970f-b4301800f774.html [Accessed: 21 June 2013].

Dornbush, J. L. (2007) "Japanese Real Estate: Does the Government Help or Hinder Development?" [online] *Cornell Real Estate Rev.*, 5, p. 4.

Dreveskracht, R. (2011) "Economic Development, Native Nations, and Solar Projects" *J. of Energy & Dev.*, 34(2), p. 136.

Dreveskracht, R. (2012) "Alternative Energy in American Indian Country: Catering to Both Sides of the Coin" *Energy L.J.*, 33, pp. 440–48.

DSIRE Solar Portal. (2013) "Net Metering" [online] Available at: www.dsireusa. org/solar/solarpolicyguide/?id=17 [Accessed: 4 October 2013].

DSIRE USA. (2014) "Renewable Electricity Production Tax Credit (PTC)" [online] Available at: http://dsireusa.org/incentives/incentive.cfm?Incentive_Code=US13F [Accessed: 6 June 2013].

Duncan, D. and Sovacool, B. K. (2011) "The Barriers to the Successful Development of Commercial Grid Connected Renewable Electricity Projects in Australia, Southeast Asia, the United Kingdom and the United States of America" *Renewable Energy L. & Pol'y Rev.*, 2, p. 295.

Duvivier, K. (2009) "Animal, Vegetable, Mineral—Wind? The Severed Wind Power Rights Conundrum" *Washburn L.J.*, 49, p. 86.

Eaton, K. (2013) "Sioux Tribes Plan Large-scale Wind Energy Project" [online] 5 July. Available at: http://news.yahoo.com/sioux-tribes-plan-large-scale-175702015.html [Accessed: 14 August 2013].

Eilperin, J. (2010) "Pentagon Objections Hold Up Oregon Wind Farm" [online] 15 April. Available at: www.washingtonpost.com/wp-dyn/content/article/2010/04/15/AR2010041503120.html [Accessed: 22 June 2012].

Eilperin, J. (2013) "White House Solar Panels Being Installed This Week" [online] 15 August. Available at: www.washingtonpost.com/blogs/post-politics/wp/2013/08/15/white-house-solar-panels-finally-being-installed/ [Accessed: 19 August 2013].

Electric Power Research Institute. (2011) *Estimating the Costs and Benefits of the Smart Grid: A Preliminary Estimate of the Investment Requirements and Resultant Benefits of a Fully Functioning Smart Grid* 1-1, Tech. Rep. No. 10225519 [online] Available at: http://ipu.msu.edu/programs/MIGrid2011/presentations/pdfs/Reference%20Material%20-%20Estimating%20the%20Costs%20and%20Benefits%20of%20the%20Smart%20Grid.pdf [Accessed: 4 October 2013].

Ellenbogan, J. M., Grace, S., Heiger-Bernays, W., Manwell, J., Mills, D. A., Sullivan, K. A., and Weisskopf, M. (2012) *Wind Turbine Health Impact Study: Report of Independent Panel Expert*. Mass.gov. [report] Available through: www.mass.gov/eea/docs/dep/energy/wind/turbine-impact-study.pdf [Accessed: 20 June 2012].

Elliott, D. L., Holladay, C., Barchet, W., Foote, H. S., and Usky, W. (1987) *Wind Energy Resource Atlas of the United States*. NASA STI/Recon Technical Report N, 87.

EMD International. (n.d.) "EMD" [online] Available at: www.emd.dk/WindPRO/WindPRO%20Modules,%20shadow [Accessed: 21 June 2012].

Endangered Species Act of 1973 (1973). United States Code Annotated, Title 16, Chapter 35, Sections 1531–1540; 50 Code of Federal Regulations 17.3; Definitions.

Engelman, A. (2011) "Against the Wind: Conflict over Wind Energy" *Envtl. L. Rep.*, 41, p. 10551.

European Wind Energy Association. (2011) "Wind Energy Frequently Asked Questions (FAQ)" [online] Available at: www.ewea.org/index.php?id=1884 [Accessed: 14 June 2012].

Financere.nrel.gov. (2005) "German Renewable Energy Sources Act (EEG) 2009— English version [Germany]" [online] Available at: https://financere.nrel.gov/finance/node/2740 [Accessed: 24 September 2013].

Flores, A. (2013) *Comments of the Colorado River Indian Tribes on the DRECP Interim Document*. (Docket No. 09-RENEW EO-01). [report] Parker, AZ: Colorado River Indian Tribes.

Fountainebleau Hotel Corp. v. Forty-Five Twenty-Five, Inc., 114 So.2d 357, 359 (Fla. Dist. Ct. App. 1959).

Fox, D.(2008) "Solar Energy Trumps Shade in California Prosecution of Tree Owner Across a Backyard Fence: When is the Environmentalism Greener on the Other Side?" [online] 18 March. Available at: www.csmonitor.com/2008/0318/p20s01-lihc.html.

Freeman, R. L. and Kass, B. (2010) "Siting Wind Energy Facilities on Private Land in Colorado: Common Legal Issues" *Colorado Lawyer*, 39(5), p. 43.

Friends of the Columbia Gorge. (n.d.) "Friends of the Columbia Gorge: Mission/Vision" [online] Available at: www.gorgefriends.org/section.php?id=9 [Accessed: 9 December 2013].

Frosch, D. (2013) "A Struggle to Balance Wind Energy with Wildlife" [online] 17 December. Available at: www.nytimes.com/2013/12/17/science/earth/a-struggle-to-balance-wind-energy-with-wildlife.html?_r=0.

Frye, P. (2009) *Developing Energy Projects on Federal Lands: Tribal Rights, Roles, Consultation, and Other Interests (a Tribal Perspective)*. 15B. [report] Rocky Mountain Mineral Law Institute.

Galanova, M. (2013) "Spain's Sunshine Toll: Row Over Proposed Solar Tax" [online] 6 October. Available at: www.bbc.co.uk/news/business-24272061 [Accessed: 8 October 2013].

Galbraith, K. (2008a) "Ice-Tossing Turbines: Myth or Hazard?" [online] December 9. Available at: http://green.blogs.nytimes.com/2008/12/09/ice-tossing-turbines-myth-or-hazard/ [Accessed: 20 June 2012].

Galbraith, K. (2008b) "Winter Cold Puts a Chill on Energy Renewables" [online] 26 December. Available at: www.nytimes.com/2008/12/26/business/26winter.html [Accessed: 20 June 2012].

Galbraith, K. (2011) "Gulf Coast Wind Farms Spring Up, As Do Worries" *New York Times*, 10 February.

Galbraith, K. (2013) "Filling the Gaps in the Flow of Renewable Energy" [online] 3 October. Available at: www.nytimes.com/2013/10/03/business/energy-environment/Filling-the-Gaps-in-the-Flow-of-Renewable-Energy.html?_r=0 [Accessed: 9 October 2013].

Gardner, J. E. and Lehr, R. L. (2012) "Wind Energy in the West: Transmission, Operations, and Market Reforms" *Nat. Resources & Env't*, 26, p. 13.

GE Energy. (2009) "Wind Energy Basics" [online] Available at: www.ge-energy.com/content/multimedia/_files/downloads/wind_energy_basics.pdf [Accessed: 17 June 2013].

Genesis Gains Consent for Wairarapa Mega-Wind Farm. (2013) [online] 4 July. Available at: www.nbr.co.nz/article/genesis-gains-consent-wairarapa-mega-wind-farm-bd-142453 [Accessed: 16 August 2013].

Gerdes, J. (2012) "Solar-Powered Microgrid Slashes Alcatraz Island's Dependence on Fossil Fuels" [online] 30 July. Available at: www.forbes.com/sites/justingerdes/2012/07/30/solar-powered-microgrid-reduces-alcatraz-islands-dependence-on-fossil-fuels/ [Accessed: 19 August 2013].

Gergacz, J. W. (1982) "Legal Aspects of Solar Energy: Statutory Approaches for Access to Sunlight" *BC Envtl. Aff. L. Rev.*, 10, p. 1.

Global Wind Energy Council. (2012) *Global Wind Report Annual Market Update*. [report]   www.gwec.net/wp-content/uploads/2012/06/Annual_report_2012_LowRes.pdf: GWEC, p. 8.

Godoy, E. (2013) "Rural Mexican Communities Protest Wind Farms" [online] 19 June. Available at: www.ipsnews.net/2013/06/rural-mexican-communities-protest-wind-farms/ [Accessed: 15 August 2013].

Goodrich, A., James, T., and Woodhouse, M. (2012) *Residential, Commercial, and Utility-Scale Photovoltaic (PV) System Prices in the United States: Current Drivers and Cost-Reduction Opportunities*. TP-6A20-53347. [report] Golden, Co: NREL.

Gov.uk. (2011) "Wind Turbine Shadow Flicker Study Published" [online] Available at: www.decc.gov.uk/en/content/cms/news/pn11_025/pn11_025.aspx [Accessed: 21 June 2012].

Greenhalf, J. (2011) "Solar Panels Go on Bradford Cathedral Roof" [online] 24 August. Available at: www.thetelegraphandargus.co.uk/news/9213579.Solar_panels_go_on_Bradford_Cathedral_roof/ [Accessed: 19 August 2013].

Griset, T. J. (2011) "Harnessing the Ocean's Power: Opportunities in Renewable Ocean Energy Resources" *Ocean & Coastal L.J.*, 16, pp. 398–99.

GTM Energy and Solar Energy Industries Association. (2012) "U.S. Solar Market Insight" [online] Available at: www.seia.org/sites/default/files/resources/ZDgLD2dxPGYIR-2012-ES.pdf [Accessed: 8 October 2013].

Harding, G., Harding, P., and Wilkins, A. (2008) "Wind Turbines, Flicker, and Photosensitive Epilepsy: Characterizing the Flashing That May Precipitate Seizures and Optimizing Guidelines to Prevent Them" *Epilepsia*, 49(6), pp. 1095–98.

Heintzelman, M. D. and Tuttle, C. M. (2012) "Values in the Wind: A Hedonic Analysis of Wind Power Facilities" *Land Economics*, 88(3), p. 571.

Hinman v. Pac. Air Lines Trans. Corp., 84 F.2d 755 (9th Cir. 1936).

Hodges, J. (2010) "An Answer to Green Energy Could Be in the Air" *NASA*. [online] Available at: www.nasa.gov/topics/technology/features/capturingwind.html [Accessed: 25 October 2013].

Hois, E. (2013) "Colored Solar Panels Address Concerns of Aesthetics, Historic Preservation" *Solarreviews.com*. [online] Available at: www.solarreviews.com/blog/colored-solar-panels-address-concerns-of-aesthetics-historic-preservation/ [Accessed: 20 August 2013].

Howard, B. C. (2012) "Turbines Ready for Takeoff" *National Geographic*. [online] Available at: http://news.nationalgeographic.com/news/energy/2012/09/pictures/120924-flying-wind-turbines/ [Accessed: 25 October 2013].

Howe, M. (2012) "Chinese Regional Government Claims Wind Energy Is 'State-Owned'" *Windpower Monthly*. [online] Available at: www.windpowermonthly.com/article/1136930/chinese-regional-government-claims-wind-energy-state-owned [Accessed: 24 June 2013].

Huber, S. and Horbaty, R. (2012) *IEA Wind Task 28: Social Acceptance of Wind Energy Projects*. [report] Liestal, Switzerland: International Energy Agency Wind Implementing Agreement, pp. 4–27. Available at: www.ieawind.org/index_page_postings/June%207%20posts/task%2028%20final%20report%202012.pdf [Accessed: 18 June 2012].

Hull, D. (2013) "California Adopts First-in-nations Energy Storage Plan" [online] 17 October. Available at: www.mercurynews.com/business/ci_24331470/california-adopts-first-nation-energy-storage-plan [Accessed: 30 October 2013].

Hulse, C. (2005) "Two 'Bridges to Nowhere' Tumble Down in Congress" *New York Times*, 17 November, p. A19.

Illinois Commerce Commission v. FERC, 2013 WL 2451766 (2013).

Indian Tribal Energy Development and Self-Determination Act. United States Code Annotated, Title 25, Chapter 37, Section 3502; 25 Code of Federal Regulations 224. Indian tribal energy resource development.

Intergovernmental Panel on Climate Change. (2002) *IPCC Technical Paper V: Climate Change and Biodiversity*. [report] p. 16. Available through: ipcc.ch www.ipcc.ch/pdf/technical-papers/climate-changes-biodiversity-en.pdf [Accessed: 13 June 2012].

Iowa Code Annotated. (1992) Chapter 564A. Access to solar energy.

Irfan, U. (2011) "Bats and Birds Face Serious Threats from Growth of Wind Energy" *New York Times*. [online] 8 August. Available at: www.nytimes. com/cwire/2011/08/08/08climatewire-bats-and-birds-face-serious-threats-from-gro-10511.html?pagewanted=all [Accessed: 11 June 2013].

Isherwood, C. (2010) "Larger Energy Saving Improvements for your Home" *Energy Saving Warehouse*. [online] Available at: www.energysavingwarehouse.co.uk/news/69/20/High-Cost-Improvements.html [Accessed: 13 June 2012].

Jaffe, M. (2013) "Rooftop Solar Net Metering Is Being Fought Across the U.S." [online] 1 September. Available at: www.denverpost.com/business/ci_23986631/rooftop-solar-net-metering-is-being-fought-across [Accessed: 8 October 2013].

Joint E&P Forum. (2014) "Environmental Management in Oil and Gas Exploration and Production: An Overview of Issues and Management Approaches" [report] London: pp. 17–20. Available through: ogp.org www.ogp.org.uk/pubs/254.pdf [Accessed: 13 June 2013].

Jordan, B. and Perlin, J. (1979) "Solar Energy Use and Litigation in Ancient Times" *Solar L. Rep.*, 1, p. 583.

Joskow, P. L. (2012) "Creating a Smarter US Electricity Grid" *The J. of Econ. Perspectives*, 26(1), p. 31.

Kallies, A. (2011) "The Impact of Electricity Market Design on Access to the Grid and Transmission Planning for Renewable Energy in Australia: Can Overseas Examples Provide Guidance?" *Renewable Energy L. & Pol'y Rev.*, 2, p. 158.

Kandt, A., Hotchkiss, E., Walker, A., Buddenborg, J., and Lindberg, J. (2011) "Implementing Solar PV Projects on Historic Buildings and in Historic Districts" [online] September. Available at: www.nrel.gov/docs/fy11osti/51297.pdf [Accessed: 20 August 2013].

Kapla, K. and Trummel, C. (2010) "Severing Wind Rights Raises Legal Issues" [online] October. Available at: http://kaplalaw.com/NAW1010_WindRightsArticle. pdf.

Kaplan, C. S. (2004) "Congress, the Courts, and the Army Corps: Siting the First Offshore Wind Farm in the United States" *BC Envtl. Aff. L. Rev.*, 31, p. 191.

Kates, W. (2009) "Wind Farms Interfering with Weather Radar in N.Y." [online] 13 October. Available at: www.usatoday.com/weather/research/2009-10-13-wind-farms-weather-radar_N.htm [Accessed: 22 June 2012].

Kearl, J. R. (1993) *Principles of Economics*. Lexington, MA.: D. C. Heath and Co. pp. 412–28.

Kelly-Detwiler, P. (2013) "Energy Storage: Continuing to Evolve" [online] 15 May. Available at: www.forbes.com/sites/peterdetwiler/2013/05/15/energy-storage-continuing-to-evolve/ [Accessed: 25 October 2013].

Kimmell, K. and Stalenhoef, D. S. (2011) "The Cape Wind Offshore Wind Energy Project: A Case Study of the Difficult Transition to Renewable Energy" *Golden Gate U. Envtl. L.J.*, 5, p. 200.

Kittle, S. (2012) "Wind Turbine Fire Under Investigation" [online] 30 January. Available at: http://pressrepublican.com/0100_news/x897046002/Wind-turbine-fire-under-investigation [Accessed: 25 June 2012].

Kiwi (NZ) to English Dictionary. (n.d.) "New Zealand A to Z" [online] Available at: www.newzealandatoz.com/index.php?pageid=357 [Accessed: 16 August 2013].

Klass, A. (2012) "Energy and Animals: A History of Conflict" *San Diego J. of Climate & Energy L.*, 3, p. 159.

Kooles, K. and Miller, J. (2012) "Installing Solar Panels on Historic Buildings" [online] August. Available at: http://ncsc.ncsu.edu/wp-content/uploads/Installing-Solar-Panels-on-Historic-Buildings_FINAL_2012.pdf [Accessed: 19 August 2013].

Krauss, C. (2008) "Move Over, Oil, There's Money in Texas Wind" *New York Times*, 23 February, p. A1.

Lagesse, D. (2013) "Mexico's Robust Wind Energy Prospects Ruffle Nearby Villages in Oaxaca" [online] 7 February. Available at: http://news.nationalgeographic.com/news/energy/2013/02/pictures/130208-mexico-wind-energy/ [Accessed: 15 August 2013].

Land-Based Wind Energy Guidelines. (2012) [report] U.S. Fish and Wildlife Service. p. vi. Available through: fws, gov www.fws.gov/windenergy/docs/weg_final.pdf [Accessed: 12 June 2013].

Lardinois, F. (2013) "Google X Acquires Makani Power and Its Airborne Wind Turbines" *TechCrunch*. [online] Available at: http://techcrunch.com/2013/05/22/google-x-acquires-makani-power-and-its-airborne-wind-turbines [Accessed: 25 October 2013].

Larson, E. (2011) "Cause of Action to Challenge Development of Wind Energy Turbine or Wind Energy Farm" *50 Causes of Action*, 2d 1, § 16.

Lawcommission.justice.gov.uk. (2013) "Rights to Light—Law Commission" [online] Available at: http://lawcommission.justice.gov.uk/consultations/rights-to-light.htm [Accessed: 25 June 2013].

Lawrence, D. (2013) "U.S. Poised to Catch the Offshore Wind" [online] 23 October. Available at: www.theglobeandmail.com/report-on-business/breakthrough/canada-us-poised-to-catch-the-offshore-wind/article14988940/ [Accessed: 23 February 2014].

Learn, S. (2010a) "Air Force Concerns About Radar Interference Stall Huge Oregon Wind Energy Farm" [online] 14 April. Available at: www.oregonlive.com/environment/index.ssf/2010/04/air_force_concerns_about_radar.html [Accessed: 22 June 2012].

Learn, S. (2010b) "Pentagon Drops Opposition to Big Oregon Wind Farm" [online] 30 April. Available at: www.oregonlive.com/environment/index.ssf/2010/04/air_forces_drops_opposition_to.html.

Lee, M. (2012) "Ocotillo Wind Project Advances Despite Tribal Objections" *San Diego Union Tribune*, 12 May.

Lee, S., Churchfield, M., Moriarty, P., Jonkman, J., and Michalakes, J. (2011) "Atmospheric and Wake Turbulence Impacts on Wind Turbine Fatigue Loading" [online] December. Available at: www.nrel.gov/docs/fy12osti/53567.pdf [Accessed: 19 June 2013].

Leiserowitz, A., Maibach, E., Roser-Renouf, C., Feinberg, G., and Howe, P. (2012) *Public Support for Climate and Energy Policies in September 2012.* [report] New Haven, CT: Yale Project on Climate Change Communication, p. 4.

Levitan, D. (2012) "High-Altitude Wind Energy: Huge Potential—and Hurdles" *Yale Environment 360*. [online] Available at: http://e360.yale.edu/feature/high_altitude_wind_energy_huge_potential_and_hurdles/2576/ [Accessed: 25 October 2013].

Levitan, D. (2013) "Will Offshore Wind Finally Take Off on U.S. East Coast?" *Yale Environment 360*. [online] Available at: http://e360.yale.edu/feature/will_offshore_wind_finally_take_off_on_us_east_coast/2693/ [Accessed: 16 October 2013].

Lindley, D. (2010) "Smart Grids: The Energy Storage Problem" *Nature*, 463, p. 18. Available at: www.nature.com/news/2010/100106/full/463018a.html.

Loc.gov. (n.d.) "Indian Removal Act: Primary Documents of American History (Virtual Programs and Services, Library of Congress)" [online] Available at: www.loc.gov/rr/program/bib/ourdocs/Indian.html [Accessed: 18 December 2013].

Lovich, J. E. and Ennen, J. R. (2011) "Wildlife Conservation and Solar Energy Development in the Desert Southwest, United States" *BioScience*, 61(12), pp. 982–92.

Lynch v. Hill, 6 A.2d 614, 618 (Del. Ch. 1939).

MacCleery, B. (2011) "The Advent of Airborne Wind Power" [online] January. Available at: http://windsystemsmag.com/article/detail/187/the-advent-of-airborne-wind-power [Accessed: 25 October 2013].

MacCourt, D. (2010) *Renewable Energy Development in Indian Country: A Handbook for Tribes*. [report] Golden, Co: NREL and AterWynne LLP.

Madrigal, M. and Stoft, S. (2011) *Transmission Expansion for Renewable Energy Scale-Up: Emerging Lessons and Recommendations*. [report] The Energy and Mining Sector Board, p. 95.

Mahdi, W. and Roca, M. (2012) "Saudi Arabia Plans $109 Billion Boost for Solar Power" *Bloomberg*. [online] Available at: www.bloomberg.com/news/2012-05-10/saudi-arabia-plans-109-billion-boost-for-solar-power.html [Accessed: 5 November 2013].

Makani Power. (2011) "Airborne Wind Energy" [online] Available at: www.makanipower.com/home/ [Accessed: 25 October 2013].

Malewitz, J. (2013) "Wind, Solar Could Benefit from Kansas Transmission Compact" [online] 29 July. Available at: www.midwestenergynews.com/2013/07/29/wind-solar-could-benefit-from-kansas-transmission-compact/ [Accessed: 19 September 2013].

Martin, S. L. (2009) "Wind Farms and NIMBYS: Generating Conflict, Reducing Litigation" *Fordham Envtl. L. Rev.*, 20, pp. 448–65.

Marvel, K., Kravitz, B., and Caldeira, K. (2012) "Geophysical Limits to Global Wind Power" *Nature Climate Change*, 3(2), p. 118.

Massachusetts General Laws Annotated. (2006) Part I, Title VII, Chapter 40A, Section 3. Administration of the government; Cities, towns and districts; zoning; subjects which zoning may not regulate; exemptions; public hearings; temporary manufactured home residences.

Masterson, C. D. (2009) "Wind-Energy Ventures in Indian Country: Fashioning a Functional Paradigm" *American Indian L. Rev.*, 34, p. 337.

McCrary, M. D., Mckernan, R. L., Schreiber, R. W., Wagner, W. D., and Sciarrotta, T. C. (1986) "Avian Mortality at a Solar Energy Power Plant" *J. of Field Ornithology*, 57(2), pp. 135–41.

McGovern, M. (2012) "Land Right Protests Hold Up Construction in Oaxaca" *Windpower Monthly*. [online] Available at: www.windpower-monthly.com/article/1161708/land-right-protests-hold-construction-oaxaca [Accessed: 16 August 2013].

McKee, S. J. (1982) "Solar Access Rights" *Urb. L. Ann.*, 23, p. 437.

McLean, I. F. (1999) "The Role of Legislation in Conserving Europe's Endangered Species" *Conservation Biology*, 13(5), p. 996.

Mendick, R. (2011) "Military Radar Deal Paves Way for More Wind Farms Across Britain" [online] 27 August. Available at: www.telegraph.co.uk/earth/energy/

windpower/8726922/Military-radar-deal-paves-way-for-more-wind-farms-across-Britain.html [Accessed: 12 December 2011].

Mendonca, M., Jacobs, D., and Sovacool, B. K. (2010) *Powering the Green Economy*. London: Earthscan.

Migratory Bird Treaty Act. (2004) United States Code Annotated, Title 16, Chapter 7, Subchapter II, Section 703; Code of Federal Regulations, Title 50, Part 10.13. Taking, killing or possessing migratory birds is unlawful.

Migratory Bird Treaty Act of 1918. (2004) United States Code Annotated, Title 16, Chapter 7, Subchapter II, Section 707. Violations and penalties; forfeitures.

Mills, A., Phadke, A., and Wiser, R. (2010) *Exploration of Resource and Transmission Expansion Decisions in the Western Renewable Energy Zone Initiative* [online] February. Available at: http://emp.lbl.gov/sites/all/files/REPORT%20lbnl-3077e.pdf [Accessed: 22 August 2013].

Monies, P. (2012) "Canadian County Crop Duster Files Lawsuit Against Developers of Planned Wind Farm in Oklahoma" [online] 25 May. Available at: http://newsok.com/canadian-county-crop-duster-files-lawsuit-against-developers-of-planned-wind-farm-in-oklahoma/article/3678347 [Accessed: 23 June 2012].

Mont. Code Ann. § 70-17-404 (2011).

Mortensen, B. (2007) "International Experiences of Wind Energy" *Envtl. & Energy L. &Pol'y J.*, 2, pp. 190–91.

Muenster, R. J. (2009) "Shade Happens" *Renewable Energy World*. [online] Available at: www.renewableenergyworld.com/rea/news/article/2009/02/shade-happens-54551 [Accessed: 25 June 2013].

Muscarello v. Ogle County Bd. of Comm'rs, 131 S.Ct. 1045 (2011).

Muscarello v. Winnebago County Bd., 702 F.3d 909, 912 (7th Cir. 2012).

Musial, W. and Ram, B. (2010) "Large-Scale Offshore Wind Power in the United States" [report] NREL. Available at: www.nrel.gov/docs/fy10osti/40745.pdf.

Nagle, J. C. (2013) "Green Harms of Green Projects" *Notre Dame J. L., Ethics & Pub. Pol'y*, 27, p. 72.

NASA's Imagine the Universe. (n.d.) "Electromagnetic Spectrum—Introduction" [online] Available at: http://imagine.gsfc.nasa.gov/docs/science/know_l1/emspectrum.html [Accessed: 25 February 2011].

National Environmental Justice Advisory Council: Indigenous People's Subcommittee. (2000) *Guide on Consultation and Collaboration with Indian Tribal Governments and Public Participation Groups and Tribal Members in Environmental Decision Making*. EPA/300-R-00-009, 2000. [report] Washington, D.C.: National Environmental Justice Advisory Council, p. 5.

National Historic Preservation Act. (2006) United States Code Annotated, Title 16, Chapter 1A, Subchapter II, Section 470. Congressional finding and declaration of policy.

National Register of Historic Places Determination of Eligibility Comment Sheet. (2010) [online] Available at: www.nps.gov/nr/publications/guidance/NantucketSoundDOE.pdf [Accessed: 18 October 2013].

Navrud, S., Ready, R. C., Magnussen, K., and Bergland, O. (2008) "Valuing the Social Benefits of Avoiding Landscape Degradation from Overhead Power Transmission Lines: Do Underground Cables Pass the Benefit–Cost Test?" *Landscape Research*, 33(3), pp. 281–96.

N.D. Cent. Code § 17-04-04 (2012).

New Mexico Solar Rights Act. (1978) Chapter 47, Article 3, Sections 1–4.

New Mexico Statutes. (2011) Chapter 3, Article 18, Sections 1–32. Municipalities; Powers; Limitation of county and municipal restrictions on solar collectors.

Nordman, E. (2010) "Wind Power and Human Health: Flicker, Noise and Air Quality" [online] August. Available at: www.miseagrant.umich.edu/downloads/research/projects/10-733-Wind-Brief2-Flicker-Noise-Air-Quality2.pdf [Accessed: 21 June 2012].

Norris, J. R. and Dennis, J. S. (2011) "Electric Transmission Infrastructure: A Key Piece of the Energy Puzzle" *Nat. Resources & Env't.*, 25, p. 5.

North Dakota Engrossed HB No. 1460 (2011).

Nrel.gov. (2013) "NREL: State and Local Activities—Renewable Portfolio Standards" [online] Available at: www.nrel.gov/tech_deployment/state_local_activities/basics_portfolio_standards.html [Accessed: 5 June 2013].

Oaxaca: Communards from Unión Hidalgo Request that the TUA Totally Nullify Demex Contracts. (2013) *SIPAC Blog* [blog] 24 June. Available at: http://sipazen.wordpress.com/tag/zapotec/ [Accessed: 16 August 2013].

O'Carroll, E. (2008) "German Scientists Develop Silencer for Wind Turbines" [online] 22 August. Available at: www.csmonitor.com/Environment/Bright-Green/2008/0822/scientists-develop-silencer-for-wind-turbines [Accessed: 7 November 2013].

Ocotillo Wind Energy Facility: Peninsular Bighorn Sheep Mitigation and Monitoring Plan. (2012) [report] La Mesa: Helix Environmental Planning. Available through: www.ocotilloeccmp.com/Wild1s_PBS_MMP.pdf [Accessed: 10 June 2013].

Office of the Director of Defense Research and Engineering. (2006) *The Effect of Windmill Farms on Military Readiness.* [report] Department of Defense, p. 52. Available at: www.defense.gov/pubs/pdfs/windfarmreport.pdf [Accessed: 28 September 2011].

Office of Energy Efficiency and Renewable Energy. (2012) "Wind Program" [online] Available at: http://www1.eere.energy.gov/wind/wind_potential.html [Accessed: 13 June 2012].

Official Journal of the European Union. (2009) "Directive 2009/28/EC of April 2009 on the Promotion of the Use of Energy from Renewable Sources and Amending and Subsequently Repealing Directives 2001/77/EC and 2003/30/EC, OJ L 140/19" [online] Available at: http://eur-lex.europa.eu/LexUriServ/LexUriServ.do?uri=Oj:L:2009:140:0016:0062:en:PDF [Accessed: 24 September 2013].

"Ontario Ojibwe Rally to Battle Wind-Farm Plans on Sacred Nor'Wester Mountains." (2013) [online] 29 May. Available at: http://indiancountrytodaymedianetwork.com/2013/05/29/ontario-ojibwe-rally-battle-wind-farm-plans-sacred-norwester-mountains-149599 [Accessed: 16 August 2013].

Orton, K. (2012) "Solar Power Project Eclipsed in D.C." [online] 21 June. Available at: http://articles.washingtonpost.com/2012-06-21/news/35461157_1_solar-panels-roof-solar-power [Accessed: 19 August 2013].

Osage Nation v. Wind Capital Group, LLC, 2011 WL 6371384 (N.D. Okla. 2011).

Otsego County Ordinances § 18.5.3.3 (2003).

Patel, S. (2013) "Corrected: Challenges to Order 1000 Filed in Federal Court as President Acts on Grid Modernization" *POWER Magazine.* [online] Available at: www.powermag.com/correctedchallenges-to-order-1000-filed-in-federal-court-as-president-acts-on-grid-modernization/ [Accessed: 24 September 2013].

Peabody Energy. (2013) "Kayenta Mine" [online] Available at: www.peabodyenergy.com/content/276/Publications/Fact-Sheets/Kayenta-Mine [Accessed: 20 August 2013].

Pentland, W. (2013) "No End in Sight for Spain's Escalating Solar Crisis" [online] 16 August. Available at: www.forbes.com/sites/williampentland/2013/08/16/no-end-in-sight-for-spains-escalating-solar-crisis/ [Accessed: 8 October 2013].

Peterson, R. (2012) "Mexico: Federal Court Halts Controversial Wind Park" [online] 27 December. Available at: http://globalvoicesonline.org/2012/12/27/mexico-federal-court-halts-controversial-wind-park/ [Accessed: 15 August 2013].

Piedmont Environmental Council v. FERC, 558 F.3d 304 (4th Cir. 2009).

Pierpont, N. (2009) *Wind Turbine Syndrome*. Santa Fe, NM: K-Selected Books. Available at: www.windturbinesyndrome.com/wind-turbine-syndrome/ [Accessed: 19 June 2012].

Platt, E. (2013) "The London Array: The World's Largest Offshore Wind Farm" *The Telegraph* [online] 28 July. Available at: www.telegraph.co.uk/earth/energy/windpower/9427156/The-London-Array-the-worlds-largest-offshore-wind-farm.html [Accessed: 15 October 2013].

Portman, M. E., Duff, J. A., Köppel, J., Reisert, J., and Higgins, M. E. (2009) "Offshore Wind Energy Development in the Exclusive Economic Zone: Legal and Policy Supports and Impediments in Germany and the US" *Energy Pol'y*, 37(9), p. 3604.

Potomac Economics. (2013) *2012 State of the Market Report for the Midwest ISO*. [online] June. Available at: www.potomaceconomics.com/uploads/reports/2012_SOM_Report_final_6-10-13.pdf

Powell, T. H. (2012) "Revisiting Federalism Concerns in the Offshore Wind Energy Industry in Light of Continued Local Opposition to the Cape Wind Project" *BUL Rev.*, 92, pp. 2034.

Powerroots. (n.d.) "Photovoltaic Systems" [online] Available at: http://powerroots.com/pr/renewables/solar_photovoltaic.cfm (United States photovoltaic resources map produced by the National Renewable Energy Laboratory) [Accessed: 13 June 2012].

Prah v. Maretti, 321 N.W.2d 182, 184 (1982).

Prosser, W. L. and Wade, J. W. (1956) *Restatement of the Law, Second, Torts*. Philadelphia: Executive Office, American Law Institute.

Quechan Tribe of the Fort Yuma Indian Reservation v. U.S. Dep't of the Interior, 755 F.Supp.2d 1104, 1106 (S.D. Cal. 2010).

Quechan Tribe of the Fort Yuma Indian Reservation v. U.S. Dep't of the Interior, 2012 WL 1857853 (S.D. Cal. May 22, 2012).

Quechan Tribe of the Fort Yuma Indian Reservation v. U.S. Dep't of the Interior, 2013 WL 755606 at 1 (S.D. Cal. Feb. 27, 2013).

Raftery, M. (2013) "Ocotillo Wind Project Advances Despite Tribal Objections" [online] 26 April. Available at: http://eastcountymagazine.org/node/13103 [Accessed: 14 August 2013].

Randall, T. (2013) "Wal-Mart Now Draws More Solar Power Than 38 U.S. States" *Bloomberg*. [online] Available at: www.bloomberg.com/news/2013-10-24/wal-mart-now-has-more-solar-than-38-u-s-states-drink-.html [Accessed: 5 November 2013].

Randazzo, R. (2013) "Commission Votes to Raise APS Solar Customers' Bills" [online] 14 November. Available at: www.azcentral.com/business/arizonaeconomy/articles/20131114aps-solar-customer-bills-higher.html [Accessed: 19 December 2013].

Rankin v. FPL Energy, LLC, 266 S.W.3d 506 (Tex. App. 2008).

Read, C. and Lynch, D. (2014) "The Fight For Downstream Wind Flow" *Law360. com.* [online] Available at: www.law360.com/articles/247122/the-fight-for-downstream-wind-flow [Accessed: 21 June 2013].

Reimer, H. M. and Snodgrass, S. A. (2009) "Tortoises, Bats, and Birds, Oh My: Protected-Species Implications for Renewable Energy Projects" *Idaho L. Rev.,* 46, p. 545.

Riviera, M. and Roach, J. (2013) "Out of Darkness: Solar Power Sheds a Little Light on Powerless Communities" [online] 11 August. Available at: www. nbcnews.com/technology/out-darkness-solar-power-sheds-little-light-powerless-communities-6C10867721 [Accessed: 12 August 2013].

Roberts, L. (2010) "Wind Turbines Should Be Painted Purple to Deter Bats, Scientists Claim" [online] 15 October. Available at: www.telegraph.co.uk/earth/earthnews/8066012/Wind-turbines-should-be-painted-purple-to-deter-bats-scientists-claim.html [Accessed: 11 June 2013].

Rose v. Chaikin, 453 A.2d 1378 (N.J. Super. Ct. Ch. Div. 1982).

Ruhl, J. (2012) "Harmonizing Commercial Wind Power and the Endangered Species Act Through Administrative Reform" *V. and L. Rev.,* 65, p. 1769.

Rule, T. A. (2009) "A Downwind View of the Cathedral: Using Rule Four to Allocate Wind Rights" *San Diego L. Rev.,* 46, p. 207.

Rule, T. A. (2010a) "Federal Stimulus Success at Windy Point" [online] 10 May. Available at: www.cannonpowergroup.com/media/news-articles/federal-stimulus-success-at-windy-point/ [Accessed: 6 December 2013].

Rule, T. A. (2010b) "Shadows on the Cathedral: Solar Access Laws in a Different Light" *U. Ill. L. Rev.,* 2010, pp. 876–88.

Rule, T. A. (2010c) "Sharing the Wind" *The Environmental Forum,* 27(5), pp. 30–33.

Rule, T. A. (2012) "Wind Rights Under Property Law: Answers Still Blowing in the Wind" *Probate and Property,* 26(6), pp. 56–59.

Rule, T. A. (2013) "Property Rights and Modern Energy" *George Mason L. Rev.,* 20(3), pp. 803–36.

Sahagun, L. (2012a) "Discovery of Indian Artifacts Complicates Genesis Solar Project" *Los Angeles Times,* 24 April.

Sahagun, L. (2012b) "Canine Distemper in Kit Foxes Spreads in Mojave Desert" *Los Angeles Times.* [online] 18 April. Available at: http://articles.latimes.com/2012/apr/18/local/la-me-0418-foxes-distemper-20120418 [Accessed: 10 June 2013].

Saint Index. (2011) "2011 Results" [online] Available at: http://saintindex.info/special-report-energy#windfarms [Accessed: 14 June 2012].

Sally, B. (2012) "Vestas Wind Turbine Catches Fire in Germany, No Injuries" [online] 30 March. Available at: www.bloomberg.com/news/2012-03-30/vestas-says-turbine-catches-fire-at-gross-eilstorf-wind-farm.html [Accessed: 25 June 2012].

Sanders, R. L. and Randazzo, R. (2012) "Arizona's Solar Energy Plans Vex Military" [online] 7 April. Available at: www.azcentral.com/news/articles/2012/03/23/2012 0323arizona-solar-energy-plans-military.html [Accessed: 22 June 2012].

Satherley, D. (2012) "Treaty Gives Us the Wind—Ngapuhi Leader" *3News.* [online] 5 September. Available at: www.3news.co.nz/Treaty-gives-us-the-wind-Ngapuhi-leader/tabid/1607/articleID/268086/Default.aspx [Accessed: 12 August 2013].

Schroeder, E. (2010) "Turning Offshore Wind On" *Cal. L. Rev.,* 98, pp. 1650–51.

Schumacher, A., Fink, S., and Porter, K. (2009) "Moving Beyond Paralysis: How

States and Regions Are Creating Innovative Transmission Policies for Renewable Energy Projects" *The Electricity J.*, 22(7), p. 27.

S.D. Codified Laws § 43-13-19 (2004).

Seelye, K. (2013) "Koch Brother Wages 12-Year Fight Over Wind Farm" *New York Times*, 22 October.

Seia.org. (2012) "Solar Means Business: Top Commercial Solar Customers in the U.S." [online] Available at: www.seia.org/research-resources/solar-means-business-top-commercial-solar-customers-us [Accessed: 30 June 2013].

Shipman, C. (2012) "Maori Water Rights Claim at Supreme Court" *3News.* [online] 5 September. Available at: www.3news.co.nz/Maori-water-rights-claim-at-Supreme-Court/tabid/1607/articleID/285150/Default.aspx [Accessed: 16 August 2013].

Shugarts, S. (2013) "AZ Solar Customers Face New Proposed Fees" *Energybiz.* [online] Available at: www.energybiz.com/article/13/09/az-solar-customers-face-new-proposed-fees [Accessed: 8 October 2013].

"Slight Change in Wind Turbine Speed Significantly Reduces Bat Mortality." (2010) [online] 3 November. Available at: www.sciencedaily.com/releases/2010/11/101101115619.htm [Accessed: 12 June 2013].

Smallwood, K. S. (2013) "Comparing Bird and Bat Fatality-rate Estimates Among North American Wind-energy Projects" *Wildlife Society Bulletin*, 37(1), pp. 19–33.

Smith, J. M. (2012) "Indigenous Communities in Mexico Fight Corporate Wind Farms" *Upsidedownworld.org.* [online] Available at: http://upsidedownworld.org/main/mexico-archives-79/3952-indigenous-communities-in-mexico-fight-corporate-wind-farms [Accessed: 16 August 2013].

Smith, N. and De Vries, E. (2004) "Wind and Fire: Reducing the Risk of Fire Damage in Wind Turbines" [online] Sept.–Oct. Available at: www.firetrace.com/wp-content/uploads/windandfirearticle.pdf [Accessed: 25 June 2012].

Sneed, D. (2010) "Giant Kangaroo Rat Puts Kink in California Valley Solar Project" [online] 11 September. Available at: www.sanluisobispo.com/2010/09/11/1284985/giant-kangaroo-rat-puts-kink-in.html [Accessed: 10 June 2013].

Snyder, J. and Johnsson, J. (2013) "Exelon Falls From Green Favor as Chief Fights Wind Aid" *Bloomberg.* [online] Available at: www.bloomberg.com/news/2013-04-01/exelon-falls-from-green-favor-as-chief-fights-wind-aid.html [Accessed: 3 October 2013].

Sowinski, M. (2011) "Practical, Legal, and Economic Barriers to Optimization in Energy Transmission and Distribution" *J. Land Use & Envtl. L.*, 26, p. 521.

Spence, D. B. (2011) "Regulation, Climate Change, and the Electric Grid" *San Diego J. Climate & Energy L.*, 3, p. 267.

Spinelli, D. (2010) "Historic Preservation and Offshore Wind Energy: Lessons Learned from the Cape Wind Saga" *Gonz. L. Rev.*, 46(30), p. 748.

Summary of Public Scoping Comments Received During the Scoping Period for the Solar Energy Development Programmatic Environmental Impact Statement. (2008) [report] Washington, D.C.: U.S. Department of Energy and Bureau of Land Management. p. 7. Available through: Solar Energy Development Programmatic EIS Information Center, http://solareis.anl.gov/documents/docs/scoping_summary_report_solar_peis_final.pdf [Accessed: 10 June 2013].

Starr, S. (2010) "Turbine Fire Protection" *Wind Systems.* [online] August. Available at: http://windsystemsmag.com/article/detail/136/turbine-fire-protection [Accessed: 25 June 2012].

Sussman, E. (2008) "Reshaping Municipal and County Laws to Foster Green Building, Energy Efficiency, and Renewable Energy" *NYU Envtl. L.J.*, 16, p. 1.

Tabuchi, H. (2013) "To Expand Offshore Power, Japan Builds Floating Windmills" *New York Times*, 24 October.

Takagi, F. (1977) "Legal Protection of Solar Rights" *Conn. L. Rev.*, 10, p. 136.

Tanana, H. J. and Ruple, J. C. (2012) "Energy Development in Indian Country: Working Within the Realm of Indian Law and Moving Towards Collaboration" *Utah Envtl. L. Rev.*, 32, p. 1.

Testa, J. (2012) "Sacred Ground? Citing 'viewshed,' Tribe Pushes Back Against Solar Plant" [online] 1 May. Available at: http://cronkitenewsonline.com/2012/05/sacred-ground-citing-significant-views-tribe-pushes-back-against-solar-plant/ [Accessed: 15 August 2013].

The Money Converter (n.d.) "Convert Mexican Peso to United States Dollar" [online] Available at: http://themoneyconverter.com/MXN/USD.aspx [Accessed: 15 August 2013].

Thomsen, K. and Sorensen, P. (1999) "Fatigue Loads for Wind Turbines Operating in Wakes" *J. Wind Engineering and Industrial Aerodynamics*, 80(1), pp. 121–36.

Tom, Z. (2012) "For Those Near, the Miserable Hum of Clean Energy" [online] 5 October. Available at: www.nytimes.com/2010/10/06/business/energy-environment/06noise.html [Accessed: 19 June 2012].

Touchette, M. (2014) "Wind Farm Opponents Cite Concerns for Crop Dusting" [online] 3 August. Available at: www.reviewatlas.com/news/x84682246/Wind-farm-opponents-site-concerns-for-crop-dusting?zc_p=0 [Accessed: 25 June 2012].

Trabish, H. K. (2013) "Rooftop Solar Scores a Crucial Win Against Arizona's Dominant Utility: Greentech Media" *Greentech Solar*. [online] Available at: www.greentechmedia.com/articles/read/Crucial-Win-For-Rooftop-Solar-Over-Arizonas-Dominant-Utility [Accessed: 19 December 2013].

United States v. Causby, 328, U.S. 256, 264 (1946).

United States Federal Energy Regulatory Commission. (2011) "Briefing on Order 1000" *The Final Rule on Transmission Planning and Cost Allocation*. [online] Available at: www.ferc.gov/media/news-releases/2011/2011-3/07-21-11-E-6-presentation.pdf [Accessed: 24 September 2013].

United States of America Federal Energy Regulatory Commission. (2014) *Transmission Planning and Cost Allocation by Transmission Owning and Operating Public Utilities*. 136 FERC ¶ 61051. [report].

U.S. Census Bureau. (2012) "Statistical Abstract of the United States: 2012" [online] Available at: www.census.gov/compendia/statab/2012/tables/12s0014.pdf [Accessed: 5 September 2013].

U.S. Department of Energy: Energy Efficiency and Renewable Energy. (n.d.) "The History of Solar" [online] Available at: http://www1.eere.energy.gov/solar/pdfs/solar_timeline.pdf [Accessed: 26 June 2013].

U.S. Department of the Interior Bureau of Land Management. (n.d.) "Renewable Energy Projects Approved Since 2009" [online] Available at: www.blm.gov/wo/st/en/prog/energy/renewable_energy/Renewable_Energy_Projects_Approved_to_Date.html [Accessed: 13 August 2013].

U.S. Fish and Wildlife Service. (n.d.) "West Virginia Field Office" *Fws.gov*. [online] Available at: www.fws.gov/westvirginiafieldoffice/beech_ridge_wind_power.html [Accessed: 12 June 2013].

Vaheesan, S. (2012) "Preempting Parochialism and Protectionism in Power" *Harvard J. on Legis.*, 49, p. 97.

Vance, E. (2012) "The 'Wind Rush': Green Energy Blows Trouble Into Mexico" *Christian Science Monitor.* [online] Available at: www.csmonitor.com/ Environment/2012/0126/The-wind-rush-Green-energy-blows-trouble-into-Mexico [Accessed: 7 November 2013].

Vestel, L. B. (2010) "Wind Turbine Projects Run Into Resistance" *New York Times*, 27 August, p. B1.

Wahl, D. and Giguere, P. (2006) "Ice Shedding and Ice Throw—Risk and Mitigation" *GE Energy.* [online] Available at: http://site.ge-energy.com/prod_serv/products/ tech_docs/en/downloads/ger4262.pdf [Accessed: 20 June 2012].

Wald, M. L. (2013) "Arizona Utility Tries Storing Solar Energy for Use in the Dark" *New York Times*, 18 October.

Wang, U. (2013) "How $5M Will Help a Startup Tackle the Emerging Energy Storage Market" [online] 24 October. Available at: www.forbes.com/sites/ uciliawang/2013/10/24/how-5m-will-help-a-startup-tackle-the-energy-storage-market/ [Accessed: 30 October 2013].

Wells, K. (2012) "Tortoises Manhandled for Solar Splits Environmentalists" *Bloomberg.* [online] Available at: www.bloomberg.com/news/2012-09-20/ tortoises-manhandled-for-solar-splits-environmentalists.html [Accessed: 13 June 2013].

Wernau, J. (2012a) "Exelon Pushes to Scrap Wind Subsidy" *Chicago Tribune*, 3 August.

Wernau, J. (2012b) "Wind Energy Group Gives Exelon the Boot" *Chicago Tribune*, 10 September.

Western Governors' Association. (2013) "Western Renewable Energy Zones" [online] Available at: www.westgov.org/component/content/article/102-initiatives/219-wrez [Accessed: 19 September 2013].

"Wildlife Slows Wind Power." (2011) *wsj.com.* [online] 10 December. Available at: http://online.wsj.com/article/SB1000142405297020350130457708859330713 2850.html [Accessed: 11 June 2013].

Willers, H. (2012) "Grounding the Cape Wind Project: How the FAA Played Into the Hands of Wind Farm Opponents and What We Can Learn From It" *J. Air L. & Com.*, 77, p. 615.

Williams, C. J. (2013) "With nuclear plants idled, Japan launches pioneering wind project" *Los Angeles Times*, 11 November.

Williams, K. P. (2001) "Cleveland Park Historic District Brochure" *District of Columbia Planning Website.* [online] Available at: http://planning. dc.gov/DC/Planning/Historic+Preservation/Maps+and+Information/ Landmarks+and+Districts/Historic+District+Brochures/Cleveland+Park+Historic +District+Brochure [Accessed: 18 August 2013].

Wilson, A. and Hirokawa, K. (2010) "Local Planning for Wind Power: Using Programmatic Environmental Impact Review to Facilitate Development" *Zoning and Planning L. Rep.*, 33(1), pp. 4–5.

Wind Directions. (2006, July/August issue) "Wind Power and the Environment—Benefits and Challenges" [online] Available at: www.ewea.org/fileadmin/ ewea_documents/documents/publications/WD/wd25-5-focus.pdf. [Accessed: 18 June 2012].

Wiseman, H. (2011) "Expanding Regional Renewable Governance" *Harv. Envtl. L. Rev.*, 35, p. 516.

Wolfe, P. (2013) "The Rise of Utility-scale Solar" *Renewable Energy World*. [online] Available at: www.renewableenergyworld.com/rea/ncws/article/2013/04/the-rise-and-rise-of-utility-scale-solar [Accessed: 2 July 2013].

Wright, G. (2012) "Facilitating Efficient Augmentation of Transmission Networks to Connect Renewable Energy Generation: The Australian Experience" *Energy Pol'y*, 44, p. 80.

Wyoming Solar Rights Act. (2011) Title 34, Article 27, Section 103.

Yakima Nation Wildlife, Range and Vegetation Resources Management Program. (n.d.) "Hunting Information for Tribal Members" [online] Available at: www.ynwildlife.org/tribalhunting.php [Accessed: 10 December 2013].

Yandle, B. (1999) "Grasping for the Heavens: 3-D Property Rights and the Global Commons" *Duke Envtl. L. & Pol'y F.*, 10, p. 13.

Ymparisto.fi. (2013) "Environment" [online] Available at: www.ymparisto.fi/en-US [Accessed: 12 June 2013].

Zillman, D., Walta, M. E., and Castiella, I. D. G. (2009) "More Than Tilting at Windmills" *Washburn L.J.*, 49, pp. 17–18.

Index